零基础学
R语言数据分析

从机器学习、数据挖掘、文本挖掘到大数据分析

李仁钟　李秋缘　编著

清华大学出版社
北京

内 容 简 介

本书共分 14 章，内容主要有 R 语言简介、数据读取与写入的方法，条件判断、循环等流程控制以及自定义函数，高级绘图、低级绘图、交互式绘图的说明，决策树、支持向量机、人工神经网络的介绍，基本统计、机器学习、数据挖掘、文本挖掘、大数据分析的应用，层次聚类法、K 平均聚类算法、模糊 C 平均聚类算法、聚类指标、基因算法及人工蜂群算法的应用。

本书适合没有程序设计经验、想要接触 R 语言的人以及对统计、机器学习、数据挖掘、文本挖掘、大数据分析有兴趣的人阅读。

本书为博硕文化股份有限公司授权出版发行的中文简体字版本
北京市版权局著作权合同登记号　图字：01-2018-1990

本书封面贴有清华大学出版社防伪标签，无标签者不得销售。
版权所有，侵权必究。举报：010-62782989，beiqinquan@tup.tsinghua.edu.cn。

图书在版编目（CIP）数据

零基础学 R 语言数据分析：从机器学习、数据挖掘、文本挖掘到大数据分析 / 李仁钟，李秋缘编著.—北京：清华大学出版社，2018（2020.11重印）
ISBN 978-7-302-51080-2

Ⅰ. ①零… Ⅱ. ①李… ②李… Ⅲ. ①程序语言—程序设计 Ⅳ. ①TP312

中国版本图书馆 CIP 数据核字（2018）第 195645 号

责任编辑：夏毓彦
封面设计：王　翔
责任校对：闫秀华
责任印制：沈　露

出版发行：清华大学出版社
　　网　　址：http://www.tup.com.cn，http://www.wqbook.com
　　地　　址：北京清华大学学研大厦 A 座　　邮　编：100084
　　社 总 机：010-62770175　　　　　　　　邮　购：010-62786544
　　投稿与读者服务：010-62776969，c-service@tup.tsinghua.edu.cn
　　质量反馈：010-62772015，zhiliang@tup.tsinghua.edu.cn
印 装 者：三河市铭诚印务有限公司
经　　销：全国新华书店
开　　本：190mm×260mm　　印　张：17.75　　字　数：454 千字
版　　次：2018 年 10 月第 1 版　　　　　　印　次：2020 年 11 月第 3 次印刷
定　　价：59.00 元

产品编号：078581-01

序

有人可能会问：近几年来，为什么 R 语言在较受欢迎的各种编程语言排行榜中逐年攀升？源于统计领域、广泛使用的 R 程序设计语言具有什么魔力吗？这是因为互联网时代带来了海量的数据，而 R 是较受欢迎的数据科学语言。如果你的工作和研究与数据科学有关，或者你现在想入门数据科学，那么推荐你学习 R 语言。

R 语言与生俱来就拥有数据统计和分析的 DNA，而且 R 语言本身并不是独立存在的程序设计语言。更准确地说，R 语言以集成在一个 R 系统或环境中的方式呈现在我们面前，这个 R 系统集数据计算、数据处理、统计分析和图形绘制等软件包于一体，是一个完整的数据科学工具软件。

如今，以互联网大数据分析为基础的人工智能，如机器学习、商业智能、数据挖掘、文本挖掘、数据可视化等领域都渴求强大、高效的数据科学工具，这种渴求让 R 大放异彩。R 系统本身就是一个开放的系统，除了传统的数据统计分析/绘图等软件包，现在更增加了机器学习、数据和文本挖掘、大数据分析等相关的诸多程序包，让 R 语言在这些领域成为"光彩夺目"的明星。

如果你对上述热门的领域之一感兴趣，并且想将 R 引入你的工作或研究中，那么本书就是一本快速参考指南。本书也可以作为完全不懂 R 软件及数据分析的读者自学 R 语言的第一本读物。本书各章提供了丰富的范例程序，因而也可以作为大专院校 R 语言的上机实践课教材。

资深架构师　赵军
2018 年 6 月

前　言

随着 R 软件的流行及普及化，许多学者和专家转而使用 R 作为研究与开发的工具。R 软件有 Windows、UNIX、Linux 及 Apple MacOS 等不同操作系统的免费版本，更有一万种以上免费程序包可供使用，所以学习 R 软件是睿智的选择。

本书内容共有 14 章，前 4 章先介绍 R 软件的基本操作和应用，第 5 章对本书所使用的程序包做完整的介绍，包含 R 软件在机器学习（Machine Learning）、数据挖掘（Data Mining）、文本挖掘（Text Mining）及大数据（Big Data）分析的相关程序包，第 6-9 章介绍各类学习算法，第 10~12 章介绍关联规则、网络社群分析及文本挖掘、图形化数据分析工具，最后两章介绍 Hadoop 和 Spark 大数据分析。

作者是福州外语外贸学院教授，本书是作者多年来从事教学的心血结晶，书中范例的程序代码丰富，可作为练习的补充材料。本书的撰写以完全不懂 R 软件及数据分析的读者为对象，对于有意愿自学的读者而言，本书也是一本不错的入门参考书。

本书配套范例程序可从下面的网址（注意区分数字和字母大小写）下载或扫描右边的二维码获取：

https://pan.baidu.com/s/17b-xnYfhICguW4wSz8pWXA

如果下载有问题，请联系 booksaga@126.com，邮件主题为"零基础学 R 语言数据分析：从机器学习、数据挖掘、文本挖掘到大数据分析"。

本书的撰写虽已力求完美，但难免会有疏漏之处，欢迎各位读者指教。

李仁钟、李秋缘
2019 年 6 月

目 录

第 1 章 R 简介 1
  1.1 开始使用 R 软件 1
  1.2 R 对象 4
    1.2.1 向量 4
    1.2.2 数组 5
    1.2.3 矩阵 7
    1.2.4 数据框 9
    1.2.5 因子 11
    1.2.6 列表 11
    1.2.7 对象转换 12

第 2 章 数据的读取与写入 14
  2.1 数据的读取 14
  2.2 数据的写入与数据集 17
  2.3 RData 格式数据的写入与读取 18
  2.4 读取 SQL Server 数据库的数据 19

第 3 章 流程控制及自定义函数 20
  3.1 条件执行 20
  3.2 循环控制 22
  3.3 自定义函数 25

第 4 章 绘图功能及基本统计 27
  4.1 高级绘图 27

4.2 低级绘图 30
4.3 交互式绘图 31
4.4 图形参数 32
4.5 基本统计 34

第5章 相关程序包的介绍 39
5.1 机器学习 39
5.2 数据挖掘 40
5.3 社交网络分析及文本挖掘 40
5.4 大数据分析 41
5.5 程序包的介绍 41

第6章 监督式学习 51
6.1 决策树 51
6.2 支持向量机 61
6.3 人工神经网络 65
6.4 组合方法 70
6.4.1 随机森林 70
6.4.2 推进法 71

第7章 无监督式学习 72
7.1 层次聚类法 72
7.2 K 平均聚类算法 75
7.3 模糊 C 平均聚类算法 77
7.4 聚类指标 83

第8章 进化式学习 86
8.1 基因算法 86
8.2 人工蜂群算法 92

第 9 章 混合式学习 95
9.1 使用 C50 和 ABCoptim 程序包范例 95
9.2 使用基因算法来调整人工神经网络参数的范例 97

第 10 章 关联规则 107
10.1 关联规则简介 107
10.2 Apriori 算法 108

第 11 章 社交网络分析和文本挖掘 117
11.1 社交网络分析 117
11.2 文本挖掘 122

第 12 章 图形化数据分析工具 125
12.1 导入数据 126
12.1.1 处理数据集 130
12.1.2 设置变量 131
12.2 探索和测试数据 131
12.3 转换数据 135
12.4 建立、评估和导出模型 137

第 13 章 大数据分析（R+Hadoop） 141
13.1 Hadoop 简介 141
13.2 R+Hadoop 142

第 14 章 SparkR 大数据分析 170
14.1 dplyr 数据处理程序包 172
14.2 SparkR 数据处理 175
14.3 SparkR 与 SQL Server 181
14.4 SparkR 与 Cassandra 184
14.5 Spark Standalone 模式 186
14.6 SparkR 数据分析 189

附录 A 下载和安装 R 197

附录 B 安装 RStudio Desktop 203

附录 C 安装 ODBC 209

附录 D 指令及用法 214

附录 E 在虚拟机上安装 R+Hadoop 218

附录 F 在虚拟机上安装 SparkR 247

参考文献 272

第 1 章 R 简介

R 是统计软件，也是一种程序设计语言。R 当初是由 Ross Ihaka 与 Robert Gentleman 开发的，类似于 AT＆T 贝尔实验室 Rick Becker、John Chambers 与 Allan Wilks 等人所开发的 S 语言。R 有 Windows、UNIX、Linux 及 Apple MacOS 等不同操作系统的版本。R 目前开发的核心团队是由世界各地不同机构所组成的，其网站位于 https://www.r-project.org，在此网站上可参阅许多有关 R 的文件、书籍及信息。R 软件的应用领域包含统计分析、数据挖掘、机器学习、推荐系统、文本挖掘及大数据分析等。

本章重点内容：

- 开始使用 R 软件
- R 对象

1.1 开始使用 R 软件

R 软件的最新版本可到网站 http://www.r-project.org 中下载，单击网页左边下载区 Download 下的 CRAN(Comprehensive R Archive Network)，再从 CRAN 中选择 CRAN Mirrors 的镜像(Mirror)网站，从中下载适合用户操作系统的最新版本。用户安装 R 软件后，也可到 http://www.rstudio.com/ 下载 RStudio。RStudio 是一个为 R 设计的集成操作软件。R 及 RStudio 在 Windows 操作系统中的安装步骤可参考附录 A 和附录 B。

R 网站中提供了功能非常强大的工具集，用户可以从 CRAN 上安装相关程序包（Package），R 提供了一万个以上免费的程序包。当用户的计算机连接到网络后，若使用 Windows 版本，则可以很容易地通过"程序包"菜单来安装程序包。用户可从该菜单中选择"加载程序包"选项来选择可用的程序包。当用户选择想要的程序包后，R 软件将下载所选择的程序包并自动进行安装。在本书中除了第 13、14 章外，主要的范例和操作都在 Windows 操作系统下进行，如果用户在 UNIX、Linux 或者 Apple MacOS 上执行 R 软件，可能需要进行微调。用户也可以自行安装程序包，例如安装 C5.0 决策树程序包"C50"（注意英文字母大小写的意义是不同的），只需要在 R 提示符 ">"后输入以下指令即可（注意：当提示符为"+"时，表示程序正在执行中，或者正在等待未完成的指令）。

```
> install.packages("C50")
```

用户可使用以下指令来使用 C50 程序包中提供的函数：

```
> library(C50)
```

若是要删除已安装的程序包，例如 C50 程序包，用户可以使用下面的指令：

```
> remove.packages("C50")
```

有些程序包无法在 CRAN "程序包"下的"加载程序包"选项中找到，此时用户需先到适当网站下载程序包的 ZIP 文件，再使用"程序包"下的"Install Package(s) from local files…"选项来安装。例如，用户要使用 ANN 程序包，需先到 http://cran.r-project.org/src/contrib/Archive/ANN/ 下载 ANN_0.1.4.zip，再选择"Install Package(s) from local files…"来安装，安装后可使用以下指令来调用 ANN 程序包中提供的函数：

```
> library(ANN)
```

如果用户想要知道计算机中已经安装了哪些程序包，可以输入：

```
> installed.packages()
```

其输出结果包含程序包版本信息等较复杂的信息。为了方便用户使用，可以使用 library() 来查看已安装程序包的简易信息：

```
> library()
```

此外，用户可以使用下面的指令来更新所有已安装的程序包：

```
> update.packages()
```

R 软件是一种语法非常简单的表达式语言（Expression Language）。R 语言通过对象（Object）来运行，这些对象使用它们的名称（Name）和值（Content，或内容）来描述其特性。对象名称（变量）的第一个字母必须为英文字母或句点"."，若以句点当对象的第一个字母，则其后接的第一个字符不能为数字，例如 .2iswrong 是不能作为对象名称的。对象不需要事先声明，但其名字的英文字母大小写代表不同的对象，因此 X 和 x 是不同的对象。R 语言保留了一些用于指令名称的保留字，如 c 与 NA 等。R 语言可使用赋值（Assignment）表达式"<-"来给对象赋值（也可以使用"="号），例如：

```
> x <- 10
> x
[1] 10
> X <- x^2
> X
[1] 100
> z <- sqrt(X)
> z
[1] 10
```

其次，也可以通过对象的数据种类（属性，Attribute）来描述对象的特性，也就是说，作用于一个对象的函数取决于其对象的属性。所有对象都有两个内在属性：数据类型（Mode）和长度（Length）。对象中的元素（Element）共有 4 种基本数据类型：数值（Numeric）、字符（Character）、复数（Complex）和逻辑（Logical），虽然也存在其他数据类型，但是并不能用来表示数据，例如函数（Function）、表达式（Expression）；长度（Length）属性用于表示对象中元素的数量。对象的数据类型和长度可以分别通过函数 mode() 和 length() 来获取。

```
> x <- 10
> x [1] 10

> mode(x)
[1] "numeric"

> length(x)
[1] 1
```

如果要在同一行内执行多条指令，那么可以用分号";"隔开 R 指令，例如：

```
> x <- 10; y <- x^2; z <- sqrt(y)
> z
[1] 10
```

注释可以放在程序中的任何地方，从"#"号开始到句子结束之间的语句就是注释，例如：

```
> x <- 10              #整数型数值
> x
[1] 10
> mode(x)
[1] "numeric"
> length(x)
[1] 1
> y <- 10.9            #实数型数值
> y
[1] 10.9
> mode(y)
[1] "numeric"
> length(y)
[1] 1
> z <- T               #逻辑
> z
[1] TRUE
> mode(z)
[1] "logical"
> length(z)
[1] 1
> a <- "Hello"         #文本
> a
[1] "Hello"
> mode(a)
[1] "character"
> length(a)
[1] 1
> z <- 4+2i            #复数
> z
```

```
[1] 4+2i
> mode(z)
[1] "complex"
> length(z)
[1] 1
```

1.2 R 对象

R 是以面向对象为主的程序设计语言，其常用对象可以是向量（Vector）、数组（Array）、矩阵（Matrix）、因子（Factor）、数据框（Data Frame）及列表（List）等。

1.2.1 向量

向量是由包含相同数据类型的元素组成的，R 程序中最简单的结构就是由一串有序数值构成的数值（Numeric）向量。假如用户要创建一个含有 6 个数值的向量 V，且其值分别为 10、5、3.1、6.4、9.2 和 21.7，则 R 程序中的指令为 c() 函数。

```
> V <- c(10, 5, 3.1, 6.4, 9.2, 21.7)
> V
[1] 10.0 5.0 3.1 6.4 9.2 21.7
> length(V)
[1] 6
> mode(V)
[1] "numeric"
```

也可以使用 assign() 函数来实现相同的功能：

```
> assign("V", c(10, 5, 3.1, 6.4, 9.2, 21.7))
> V
[1] 10.0 5.0 3.1 6.4 9.2 21.7
> length(V)
[1] 6
> mode(V)
[1] "numeric"
```

在某些情况下，向量的元素可能会遗失。当向量中的元素为缺失值（Missing Value）时，其相关位置可给予一个特定的值 NA（需为大写）。

```
> V <- c(10, 5, NA, 6.4, 9.2, 21.7)
> V
[1] 10.0 5.0 NA 6.4 9.2 21.7
```

用户可以使用中括号 "[]" 来存取向量中的特定元素。值得注意的是，R 程序向量对象默认的第一个元素的序号（Index，也称为下标或者索引）是 1 而不是 0。

```
> V[2]
[1] 5
```

R 还提供了 Inf、–Inf 及 NaN（Not a Number），而 NULL 是指对象的长度是 0。

```
> V <- c(1,-2,0)
> V/0
[1]  Inf -Inf  NaN

> V <- NULL
> length(V)
[1] 0
```

用户也可使用冒号":"创建向量。

```
> V2=1:10
> V2
 [1]  1  2  3  4  5  6  7  8  9 10

> V2[1]
[1] 1

> V2[2:4]
[1] 2 3 4
```

1.2.2 数组

数组可以看作是多维的向量。例如，一个 3 维的数组 X 可以用 X[i,j,k]来指向特定元素。假设数组 X 的维度向量是 c(3,4,2)，则 X 中有 3×4×2=24 个元素，依次为 X[1,1,1],X[2,1,1], ... ,X[2,4,2],X[3,4,2]。

假设 X 是一个包含 24 个元素的向量：

```
> X <- 1:24
```

用户可以使用 dim()函数指定其数组维数（Dimension），让 X 变成一个 3×4×2 的 3 维数组，而 R 程序会按照行的方式排列：

```
> dim(X) <- c(3,4,2)
> X
, , 1

     [,1] [,2] [,3] [,4]
[1,]    1    4    7   10
[2,]    2    5    8   11
[3,]    3    6    9   12

, , 2

     [,1] [,2] [,3] [,4]
[1,]   13   16   19   22
[2,]   14   17   20   23
[3,]   15   18   21   24
```

用户可以让 X 变成一个 4×6 的 2 维数组：

```
> dim(X) <- c(4,6)
> X
     [,1] [,2] [,3] [,4] [,5] [,6]
[1,]   1    5    9   13   17   21
[2,]   2    6   10   14   18   22
[3,]   3    7   11   15   19   23
[4,]   4    8   12   16   20   24
```

要创建一个数组，也可以直接调用 array() 函数来创建，此函数第一个参数指定数据向量，第二个参数指定数组维数。假设要创建一个 3×4×2 的 3 维数组：

```
> X <- array(1:24, dim = c(3,4,2))
> X
, , 1

     [,1] [,2] [,3] [,4]
[1,]   1    4    7   10
[2,]   2    5    8   11
[3,]   3    6    9   12

, , 2

     [,1] [,2] [,3] [,4]
[1,]  13   16   19   22
[2,]  14   17   20   23
[3,]  15   18   21   24
```

假设要创建一个 4×6 的 2 维数组：

```
> X <- array(1:24, dim = c(4,6))
> X
     [,1] [,2] [,3] [,4] [,5] [,6]
[1,]   1    5    9   13   17   21
[2,]   2    6   10   14   18   22
[3,]   3    7   11   15   19   23
[4,]   4    8   12   16   20   24
```

值得注意的是，以下指令会创建一个所有元素都是 0 的 3 维数组：

```
> X <- array(0, dim = c(3,4,2))
> X
, , 1

     [,1] [,2] [,3] [,4]
[1,]   0    0    0    0
[2,]   0    0    0    0
[3,]   0    0    0    0

, , 2
```

```
     [,1] [,2] [,3] [,4]
[1,]  0    0    0    0
[2,]  0    0    0    0
[3,]  0    0    0    0
```

也可以使用 rbind() 和 cbind() 函数来创建数组。rbind() 表示按向量行（Row）合并成一个数组，而 cbind() 是使用列的方式合并：

```
> X1 <- c(1,2,3,4)
> X2 <- c(5,6,7,8)
> X  <- rbind(X1,X2)
> X
   [,1] [,2] [,3] [,4]
X1  1    2    3    4
X2  5    6    7    8
> X <- cbind(X1,X2)
> X
     X1 X2
[1,]  1  5
[2,]  2  6
[3,]  3  7
[4,]  4  8
```

1.2.3 矩阵

矩阵（Matrix）就是一个 2 维数组，要创建一个矩阵，可以使用函数 matrix()：

```
matrix(data = NA, nrow = 1, ncol = 1, byrow = FALSE,
dimnames = NULL)
```

其中：

byrow 表示矩阵数据是按行还是按列（byrow = FALSE）的顺序排列。
nrow 表示矩阵的行数。
ncol 表示矩阵的列数。
dimnames 表示可以帮行列命名。

```
> X <- matrix(1:24, nrow=4, ncol=6, byrow=TRUE)
> X
     [,1] [,2] [,3] [,4] [,5] [,6]
[1,]   1    2    3    4    5    6
[2,]   7    8    9   10   11   12
[3,]  13   14   15   16   17   18
[4,]  19   20   21   22   23   24

> X <- matrix(1:24, nrow=4, ncol=6, byrow=FALSE)
> X
     [,1] [,2] [,3] [,4] [,5] [,6]
```

```
[1,]   1   5    9   13   17   21
[2,]   2   6   10   14   18   22
[3,]   3   7   11   15   19   23
[4,]   4   8   12   16   20   24
```

也可以使用 rbind() 和 cbind() 函数来创建矩阵。t()是矩阵的转置（Transposition）函数，nrow()和 ncol() 函数分别返回矩阵的行数和列数。

```
> X1 <- c(1,2,3)
> X2 <- c(4,5,6)
> X3 <- c(7,8,9)
> X <- cbind(X1,X2,X3)
> X
     X1 X2 X3
[1,]  1  4  7
[2,]  2  5  8
[3,]  3  6  9
> Y=t(X)
> Y
   [,1] [,2] [,3]
X1   1    2    3
X2   4    5    6
X3   7    8    9

> m <- nrow(Y)
> m
[1] 3
> n <- ncol(Y)
> n
[1] 3
```

若要显示矩阵 X 第一行元素，则可使用：

```
> X[,1]
[1] 1 2 3
```

若要显示矩阵 X 第二列元素，则可使用：

```
> X[2,]
X1 X2 X3
 2  5  8
```

若要显示矩阵 X 第一列和第三列元素，则可使用：

```
> X[c(1,3),]
     X1 X2 X3
[1,]  1  4  7
[2,]  3  6  9
```

若要删除矩阵 X 第一行元素，则可使用：

```
> X[,-1]
     X2 X3
[1,]  4  7
[2,]  5  8
```

```
[3,]  6  9
```

若要删除矩阵 X 第二列元素，则可使用：

```
> X[-2,]
     X1 X2 X3
[1,]  1  4  7
[2,]  3  6  9
```

eigen() 函数用来计算矩阵的特征值（Eigen Value）和特征向量（Eigen Vector）：

```
> eigen(Y)
$values
[1]  1.611684e+01 -1.116844e+00 -1.303678e-15

$vectors
           [,1]        [,2]       [,3]
[1,] -0.2319707 -0.78583024  0.4082483
[2,] -0.5253221 -0.08675134 -0.8164966
[3,] -0.8186735  0.61232756  0.4082483
```

可用 %*% 表达式表示矩阵的相乘：

```
> z <- Y%*%X
> z
   X1  X2  X3
X1 14  32  50
X2 32  77 122
X3 50 122 194
```

若要修改矩阵 z 的列名称，则可使用：

```
> colnames(z) <- c("c1","c2","c3")
> z
   c1  c2  c3
X1 14  32  50
X2 32  77 122
X3 50 122 194
```

若要修改矩阵 z 的行名称，则可使用：

```
> rownames(z) <- c("r1","r2","r3")
> z
   c1  c2  c3
r1 14  32  50
r2 32  77 122
r3 50 122 194
```

1.2.4 数据框

数据框与矩阵的结构类似，因为两者的结构都是 2 维的。然而，与矩阵不同的是，数据框可以在不同行中存在不同的数据类型，但同行的数据类型和长度必须相同。数据框的每一行可视为一组观察值(Observation)或案例(Case),其变量名称是由每一列的名称来定义的。

可使用下列方式创建数据框:

```
> id <- c(1, 2, 3, 4)
> age <- c(25, 30, 35, 40)
> sex <- c("Male", "Male", "Female", "Female")
> pay <-c (30000, 40000, 45000, 50000)
> X.dataframe <- data.frame(id, age, sex, pay)
> X.dataframe
  id age    sex   pay
1  1  25   Male 30000
2  2  30   Male 40000
3  3  35 Female 45000
4  4  40 Female 50000
```

可使用下列方式获取或引用数据框中某一位置的元素:

```
> X.dataframe[3,2]
[1] 35
```

可使用列的名称获取或引用数据框中对应列的所有元素:

```
> X.dataframe$age
[1] 25 30 35 40
```

可使用下列方式获取或引用数据框中对应列的名称及元素:

```
> X.dataframe[2] age
1  25
2  30
3  35
4  40
```

R 程序提供了与 Excel 界面类似的编辑器来创建或修改数据框的值(见图 1-1):

```
> edit(X.dataframe)
```

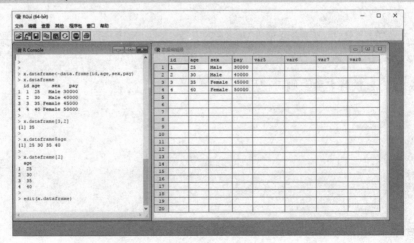

图 1-1 edit() 函数

若确定要更改"修改后"的数据框,则需使用赋值运算:

```
> X.dataframe <- edit(X.dataframe)
```

注意:必须先双击要修改或编辑的字段,才能够修改或编辑字段值。

1.2.5 因子

因子(Factor)是一种特别的向量,用于对同样长度的离散数据向量进行分组(Grouping)。文本向量中每一个元素取一个离散值,因子有一个特殊属性层级(Levels),用来表示这个分组所有的离散值。一旦设置为因子,R 打印时就不会加上双引号。

可以使用 factor() 函数来创建因子,其中的 Levels 代表分组的结果:

```
> sex <- factor(c("男","女","男","男","女"))
> sex
[1] 男 女 男 男 女
Levels: 女 男
```

1.2.6 列表

R 程序的列表是一个以对象的有序集合(Order Sequence)所构成的对象。列表中的组成元素(Component)可以是异构(Heterogeneous)的对象,也就是说,每一个组成元素的数据类型可以不相同。一个列表中的组成元素可以包括数值、逻辑、文本、复数、向量、矩阵、因子及数据框等。

可以使用 list() 函数来创建列表:

```
> id <- c(1, 2, 3)
> sex <- c("Male", "Male", "Female")
> pay <-c (30000, 40000, 45000)
> Y.dataframe <- data.frame(id, sex, pay)

> gender <- factor(c("男","男","女"))
> Paul.Family <- list(name="Paul",wife="Iris",no.kids=3, kids.age=c(25,28,30),gender, Y.dataframe)
> Paul.Family
$name
[1] "Paul"

$wife
[1] "Iris"

$no.kids
[1] 3

$kids.age
[1] 25 28 30
```

```
[[5]]
[1] 男 男 女
Levels: 女 男

[[6]]
  id    sex   pay
1  1   Male 30000
2  2   Male 40000
3  3 Female 45000
```

可以使用 $ 来获取或引用列表中某一位置的组成元素，例如要取得 Paul.Family 中第 4 个元素，语句如下：

```
> Paul.Family$kids.age
[1] 25 28 30
```

也可以使用双重中括号"[[]]"及序号值来获取或引用列表中某一位置的组成元素，例如获取 Paul.Family 中第 4 个元素，语句如下：

```
> Paul.Family[[4]]
[1] 25 28 30
```

使用一个中括号"[]"及序号值可取得或引用列表中某一位置的组成元素及名称，语句如下：

```
> Paul.Family[4]
$kids.age
[1] 25 28 30
```

要注意的是需使用双重中括号的情况（下面第二个例子）。若要获取第二个孩子的年龄，则可使用：

```
> Paul.Family$kids.age[2]
[1] 28
```

或

```
> Paul.Family[[4]][2]
[1] 28
```

1.2.7 对象转换

R 程序中提供了多个函数用于不同对象之间的转换，包括 as.vector()、as.array()、as.matrix()、as.factor()、as.data.frame() 及 as.list()函数。

创建数据框对象：

```
> id <- c(1, 2, 3, 4)
> x <- data.frame(id)
> x
  id
1  1
2  2
```

```
3  3
4  4
```

把数据框对象转换为矩阵对象:

```
> matrix.x=as.matrix(x)
> matrix.x
     id
[1,]  1
[2,]  2
[3,]  3
[4,]  4
```

把矩阵对象转换为向量对象:

```
> vector.x=as.vector(matrix.x)
> vector.x
[1] 1 2 3 4
```

第 2 章 数据的读取与写入

对于数据的读取和写入工作,可以先使用 setwd("C:/data") 或 setwd("/home/R") 函数来改变工作目录,再使用 getwd() 函数来确认当前工作目录:

```
> setwd("c:/")
> getwd()
1] "c:/"
```

本章重点内容:

- 数据的读取
- 数据的写入与数据集
- RData 格式数据的写入与读取
- 读取 SQL Server 数据库的数据

2.1 数据的读取

R 程序常用 read.table() 或 scan() 函数读取保存在文本文件(ASCII)中的数据,read.table() 函数主要是在数据框中操作,可以直接把整个外部文件读入数据框对象。scan() 函数可以直接接收键盘输入的数据。

外部文件常要求有特定的格式,例如:

(1)第 1 排(Line)可以是 header,即各列数据的变量名称(Variable Name),header 也可以省略。
(2)其余各排是各列变量名称的值。

R 程序工作目录中的 X.csv 文件如表 2-1 所示。

表 2-1 X.csv 文件

id	age	sex	pay
1	25	Male	30000
2	30	Male	40000
3	35	Female	45000
4	49	Female	50000

```
> setwd("c:/")
> X <- read.table("X.csv",sep=",",header=TRUE, encoding="GBK")
> X
  id age    sex   pay
1  1  25   Male 30000
2  2  30   Male 40000
3  3  35 Female 45000
4  4  49 Female 50000

> X$age
[1] 25 30 35 49

> X[1,2]
[1] 25
```

注意，CSV 文件是用逗点分隔字段的，所以加入 sep="," 来指定分隔符是逗点。若 header=FALSE，则使用默认的 V1,V2,…,V# 作为 header 的名称。

```
> setwd("c:/")
> X <- read.table("X.csv",sep=",", encoding="GBK")
> X
  V1  V2     V3    V4
1 id age    sex   pay
2  1  25   Male 30000
3  2  30   Male 40000
4  3  35 Female 45000
5  4  49 Female 50000
```

用户也可使用 read.csv() 函数：

```
> setwd("c:/")
> X <- read.csv("X.csv", header=TRUE, encoding="GBK")
> X
  id age    sex   pay
1  1  25   Male 30000
2  2  30   Male 40000
3  3  35 Female 45000
4  4  49 Female 50000

> X <- read.csv("X.csv", header=FALSE, encoding="UTF-8")
> X
  V1  V2     V3    V4
1 id age    sex   pay
2  1  25   Male 30000
3  2  30   Male 40000
4  3  35 Female 45000
5  4  49 Female 50000
```

用户可以使用 Excel 将 X.csv 转成 X.txt 文件（使用文本文件格式，而不是使用 Unicode 文本格式），再将文件读入。若文件中有中文，则需先确认文件格式，进行文件转码后再读文件。MacOS 操作系统的默认格式是 UTF-8，而 Windows 操作系统使用 GBK（简体中文）。使用 Windows 操作系统时，用户可使用 Notepad 打开 CSV 文件，另存为 UTF-8 格式的文件后再读入文件。

读取范例文本文件 "6-15 学生平均体重.csv"：

```
> X <- read.csv("6-15 学生平均体重.csv", header = TRUE, encoding = "UTF-8")
> X

> X <- read.table("X.txt",header=TRUE, encoding="UTF-8")
> X
  id age    sex   pay
1  1  25   Male 30000
2  2  30   Male 40000
3  3  35 Female 45000
4  4  49 Female 50000
```

scan() 函数比 read.table() 函数更加灵活，因为 scan() 函数可接收键盘输入的数据：

```
> X <- scan("")
1: 12              # 输入值后按 Enter 键
2: 10
3: 5
4: 6.3
5:                 # 不输入数据时可再按 Enter 键结束
Read 4 items
> X
[1] 12.0 10.0 5.0 6.3
```

scan() 函数也可以指定输入数据的数据类型，例如要创建列表对象：

```
> my=scan(file="",what=list(name="",pay=integer(0),sex=""))
1: peter 50000 M      # 输入值后按 Enter 键
2: lisa 40000 F
3: johnson 65000 M
4:                    # 不输入数据时可再按 Enter 键结束
Read 3 records

> mode(my) [1] "list"
```

其中：

file 文件路径，file="" 表示从键盘输入值。

what 设置输入值的数据类型，上述例子为创建列表对象，其第一个组成元素 name="" 表示文本，第二个组成元素 pay=integer(0) 表示整数，第三个组成元素 sex="" 也是文本。

scan()函数也可以读取 CSV 和文本文件，表 2-2 所示为 X1.csv 文件。

表 2-2　X1.csv 文件

id	age	pay
1	25	30000
2	30	40000
3	35	45000
4	49	50000

```
> X <- scan("X1.csv", sep=",")
Read 12 items
> X
 [1]     1    25 30000     2    30 40000     3    35 45000     4    49 50000
```

使用 Excel 将 X1.csv 转成 X1.txt 文件后再读入：

```
> X <- scan("X1.txt")
Read 12 items
> X
 [1]     1    25 30000     2    30 40000     3    35 45000     4    49 50000
```

2.2 数据的写入与数据集

若需存储数据或把分析结果输出到外部文件，则可使用 write.table() 函数：

```
> write.table(X,"C:/X_File.csv",row.names=FALSE,col.names=TRUE,sep=",",
+             fileEncoding="GBK")
```

其中：

X 表示要输出到外部文件的对象。

"C:/X_File.csv" 表示要输出到外部的文件的文件名及路径。

row.names 表示输出到外部的文件是否加上行名称。

col.names 表示输出到外部的文件是否加上列名称。

sep="," 表示分隔符。

fileEncoding 表示输出到外部的文件的格式。

R 程序提供了一些内建的数据集，可使用 data() 函数来查询已创建的数据集，如图 2-1 所示。

```
> data()
```

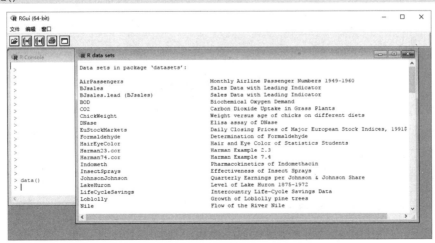

图 2-1 使用 data() 函数查询创建的数据集

可通过调用 data（数据集名称）函数来使用内建的数据集，例如要使用 iris 数据集，如图 2-2 所示。

```
> data(iris)
> iris
```

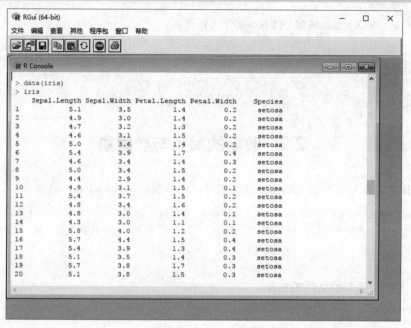

图 2-2　使用 iris 数据集

用户可调用 str() 函数来获取数据结构：

```
> str(iris)
'data.frame': 150 obs. of  5 variables:
 $ Sepal.Length: num  5.1 4.9 4.7 4.6 5 5.4 4.6 5 4.4 4.9 ...
 $ Sepal.Width : num  3.5 3 3.2 3.1 3.6 3.9 3.4 3.4 2.9 3.1 ...
 $ Petal.Length: num  1.4 1.4 1.3 1.5 1.4 1.7 1.4 1.5 1.4 1.5 ...
 $ Petal.Width : num  0.2 0.2 0.2 0.2 0.2 0.4 0.3 0.2 0.2 0.1 ...
 $ Species     : Factor w/ 3 levels "setosa","versicolor",..: 1 1 1 1 1 1 1 1 1 1 ...
```

2.3　RData 格式数据的写入与读取

R 程序可将使用的对象存储成 RData 格式的外部文件，并可读回 R 中，若用户要把 iris 数据集存储到 C:\iris.RData 中，则可使用 save() 函数。

```
> setwd("c:/")
> data(iris)
> save(iris,file="iris.RData")
```

用户若要读取 iris.RData，则可使用 load() 函数：

```
> getwd()
```

```
[1] "c:/"
> load("iris.RData", .GlobalEnv)
```

其中：

- .GlobalEnv 表示用户正在使用的工作空间（Workspace）。

2.4 读取 SQL Server 数据库的数据

首先安装 RODBC 程序包并使用 RODBC 程序包。

```
> install.packages("RODBC")
> library("RODBC")
```

用户可先将本书的 IRIS_Data 数据库附加到 MS SQL Server 中（注意要将防火墙端口打开并在 SQL Server 配置管理器中启用 TCP/IP）。用户在 Windows 操作系统环境中可使用 RODBC 程序包读取 Microsoft SQL Server 数据库的数据。

用户也可以在 CentOS 操作系统环境中使用 RODBC 程序包读取 Microsoft SQL Server 数据库的数据，可参考附录 C 有关设置 ODBC（Open Database Connectivity）的内容。

使用 odbcConnect() 函数连接 IRIS_Data 数据库：

```
> db <- odbcConnect(dsn="test", uid="test", pwd="test")
> sqlTables(db)
```

使用 sqlQuery() 读取 iris 表格（Table）内的数据：

```
> df <- sqlQuery(db, "select * from iris")
```

显示前 6 项数据：

```
> head(df)
  sepal_length sepal_width petal_length petal_width species
1          5.1         3.5          1.4         0.2  setosa
2          4.9         3.0          1.4         0.2  setosa
3          4.7         3.2          1.3         0.2  setosa
4          4.6         3.1          1.5         0.2  setosa
5          5.0         3.6          1.4         0.2  setosa
6          5.4         3.9          1.7         0.4  setosa

> str(df)
'data.frame': 150 obs. of  5 variables:
 $ sepal_length: num  5.1 4.9 4.7 4.6 5 5.4 4.6 5 4.4 4.9 ...
 $ sepal_width : num  3.5 3 3.2 3.1 3.6 3.9 3.4 3.4 2.9 3.1 ...
 $ petal_length: num  1.4 1.4 1.3 1.5 1.4 1.7 1.4 1.5 1.4 1.5 ...
 $ petal_width : num  0.2 0.2 0.2 0.2 0.2 0.4 0.3 0.2 0.2 0.1 ...
 $ species     : Factor w/ 3 levels "setosa","versicolor",..: 1 1 1 1 1 1 1 1 1 1 ...
```

结束 ODBC：

```
> odbcClose(db)
```

第3章 流程控制及自定义函数

R 程序是一种表达式语言（Expression Language），其所有的语句（或称为指令）都是函数或表达式，而赋值表达式 "<-" 的返回值就是被赋值的对象，在 R 程序中最简单的执行方式就是一排排地输入表达式并显示执行结果，例如：

```
> a <- c(1,2,3)
> x <- a+2
> x
[1] 3 4 5
```

若要直接显示执行结果，则可将语句用括号 "()" 括起来，例如：

```
> a <- c(1,2,3)
> (x <- a+2)
[1] 3 4 5
```

语句可以用大括号括起来，格式为{expr#1;…;expr#m}，以执行多条语句或指令，例如：

```
> {a <- c(1,2,3);x=a+2}
> x
[1] 3 4 5
```

R 程序的流程控制提供了条件执行（Condition Execution）与循环（Loop）等结构语句。

本章重点内容：

- 条件执行
- 循环控制
- 自定义函数

3.1 条件执行

R 语言的条件执行包含 if-else 语句、ifelse()函数、switch()函数。R 语言的 if-else 语法为：

```
if (condition) expr#1 else expr#2
```

或

```
if (condition) expr#1
```

其中：

- Condition　条件判断表达式，必须返回一个布尔值（TRUE 或 FALSE），&&（AND）与 ||（OR）常用于条件判断表达式的条件控制部分。
- expr#1　一般的表达式。
- expr#2　一般的表达式。

```
> x <- 6
> if (x>5) y=2 else y=4
> y
[1] 2

> X <- 3
> if (X<5) Y=10
> Y
[1] 10
```

若有多个表达式，则可使用大括号括起来，格式为{expr#2;…;expr#m}：

```
> X <- 3
> Y <- 1
> if (X<5 && Y<5)
+ {Y <- 10; Z <- 5}
> Y
[1] 10
> Z
[1] 5
```

R 语言的 ifelse()函数可用于简单的二分类判断，若 condition 判断为 TRUE，则返回 a；否则返回 b。其语法和范例如下：

```
ifelse (condition, a, b)
```

```
> X <- 20
> Y=ifelse(X>5, 2, 3)
> Y
[1] 2
```

R 语言的 switch()是一个函数，其语法为：

```
switch (condition, expr#1,…,expr#m)
```

其中：

- condition　可为正整数或文本。若其值为正整数 n，则执行表达式 expr#n，若 n 值大于 m 或小于 1，则 switch() 函数无返回值；若 condition 值为文本，则执行相对应的表达式。

```
> X <- 1
> switch(X, 5, sum(1:10), rnorm(5))
[1] 5

> X <- 2
> switch(X, 5, sum(1:10), rnorm(5))
```

```
[1] 55

> X <- 3
> switch(X, 5, sum(1:10), rnorm(5))
[1] -0.185252822 -0.351313575 -0.008195255 -1.920097610 -0.680803488

> X <- 4
> switch(X, 5, sum(1:10), rnorm(5))    #无返回值
>

> Y <- 1
> switch(Y, juice="Apple", meat="Pork")
[1] "Apple"
```

Switch() 函数也可使用文本，例如：

```
> Y <- "juice"
> switch(Y, juice="Apple", meat="Pork")
[1] "Apple"
```

3.2 循环控制

R 语言的循环控制包含 for、while 和 repeat，在循环中可使用 break 跳出循环体，或者使用 next 跳过当前一轮循环剩下尚未执行的语句，直接进入下一轮循环。

R 语言的 for 语句的语法为：

```
for (index in expr#1) expr#2
```

或

```
for (index in expr#1) {expr#2;…;expr#m}
```

其中：

- index   循环序号。
- expr#1   数值或文本向量，例如 1:5 或 c("A","O","B","AB")。
- expr#2   根据 index 而设计的程序块表达式。for 循环会将 expr#1 向量中的每个元素按照顺序以一次一个的方式赋值给 index，每给 index 赋值一次，就会执行一次对应的 expr#2 表达式。
- {expr#2;...;expr#m}   多个表达式。

例如：

```
> X <- 0
> for(i in 1:5) X <- X+i
> X
[1] 15

> X <- 0
> Y <- 0
```

```
> for(i in 1:5) { X<- X+i; Y <- i^2}
> X
[1] 15
> Y
[1] 25
```

R 语言的 while 循环语句的语法为：

```
while (condition) expr#1
```

或

```
while (condition) {expr#1;...;expr#m}
```

其中：

- condition　当 condition 的值为 TRUE 时，执行循环体内的表达式，并重复执行，直到 condition 的值为 FALSE 时才停止。
- expr#1　一般表达式。
- {expr#1;...;expr#m}　多个表达式。

例如，求 1+2+...+9+10=55：

```
> sum <- 0
> i <- 1
> while (i <= 10) {sum <- sum + i; i <- i + 1}
> sum
[1] 55
```

repeat 重复执行循环体内的表达式（或语句），通常在循环中设置检查循环控制的条件并与 break 并用。break 可以用于结束循环，也是结束 repeat 循环的唯一办法。

R 语言的 repeat 循环语句的语法为：

```
repeat expr
```

其中：

- expr　为一个用大括号括住的程序块（表达式或者语句的集合），必须设置循环控制条件的检查语句，若符合特定循环控制条件，则使用 break 结束循环。

例如，求 1+2+...+9+10=55：

```
> sum <- 0
> i <- 1
> repeat {
+ sum <- sum + i
+ i <- i + 1
+ if ( i > 10 ) break
+ }
> sum
[1] 55
```

关键词 break 可用于结束循环，也是结束 repeat 循环的唯一办法；而关键词 next 可以用来结束当前一轮的循环，然后进入下一轮循环。

例如，求 1+3+...+47+49=625：

```
> sum <- 0
> for (i in 1:50)
+ {
+ if ( i %% 2 == 0 ) next   # %% 是求偶数
+ sum <- sum + i   # 若i是偶数，则不执行 sum <- sum + i
+ }
> sum
[1] 625
```

R 语言中经常要用到循环，但其效率较差，所以应尽量避免使用。因此，在 R 语言中，有些函数（如 apply()、lapply() 及 sapply()等）可以更有效率地执行类似循环的语句。

apply(x, MARGIN, FUN, ...) 主要是对数组或矩阵的每一行或每一列执行一个指定函数。

其中：

- X 为所要参与计算的数组或矩阵。
- MARGIN 其值等于1或2。1表示行，2表示列。
- FUN 要执行的指定函数。

例如，调用 sum 函数求出数组每一行的总数：

```
> X <- array(1:24, dim = c(4,6))
> X
     [,1] [,2] [,3] [,4] [,5] [,6]
[1,]    1    5    9   13   17   21
[2,]    2    6   10   14   18   22
[3,]    3    7   11   15   19   23
[4,]    4    8   12   16   20   24
> apply(X,1,sum)
[1] 66 72 78 84
```

例如，调用 sum 函数求出数组每一列的总数：

```
> X <- array(1:24, dim = c(4,6))
> X
     [,1] [,2] [,3] [,4] [,5] [,6]
[1,]    1    5    9   13   17   21
[2,]    2    6   10   14   18   22
[3,]    3    7   11   15   19   23
[4,]    4    8   12   16   20   24
> apply(X,2,sum)
[1] 10 26 42 58 74 90
```

lapply(X,FUN,...) 对一个列表的每一个元素执行同一个函数，并返回函数执行的结果。

例如：

```
> X <- list(a=1:10, b=exp(-1:1))
```

```
> lapply(X,sum)
$a
[1] 55

$b
[1] 4.086161
```

sapply(X,FUN,...)的功能与 lapply(X,FUN,...) 函数类似，但其返回一个向量或矩阵对象。

```
> X <- list(a=1:10, b=exp(-1:1))
> sapply(X,sum)
        a        b
55.000000 4.086161
```

3.3 自定义函数

R 语言提供的常用函数可参考附录 D。用户也可以自定义函数，其定义如下：

```
> myfun <- function(arg#1,arg#2,...) {expr#1;…;expr#m}
```

其中，各项参数说明如下。

- arg#1,arg#2,...：输入自变量（Argument），自变量可以是一个以上。
- expr#1;…;expr#m：表达式。自定义函数可以不返回函数值，R 语言默认将函数内的最后一个表达式当作返回值，也可以使用 return() 函数返回值。

例如：

```
> X <- 1
> myfun <- function(X) { Y <- X+2; return (Y) }
> myfun(X)
[1] 3
```

R 语言的自定义函数允许自变量有默认值。若调用函数时没有输入自变量，则以默认值作为自变量的传入值；若输入自变量的传入值与默认值不同，则不使用默认值。

例如：

```
> X <- 2  # 输入自变量的传入值与默认值不同
> myfun <- function(X=1) { Y <- X+2; return (Y) }
> myfun(X)
[1] 4

> myfun <- function(X=1) { Y <- X+2; return (Y) }
> myfun()       # 若没有输入自变量，则使用默认值 X=1
[1] 3
```

在 R 语言的自定义函数中，若要修改函数外部对象的值，则只有使用 "<<-" 才能改变。
例如：

```
> x <- 1
```

```
> myfun <- function(x) { x <- 2; print(x) }
> myfun(x)
[1] 2       # myfun中 x 的值
> x         # 无法改变外部对象 x 的值
[1] 1

> x <- 1
> myfun <- function(x) { x <<- 2; print(x) }
> myfun()
[1] 2       # myfun 中改变外部对象 x 的值
> x         # 外部对象 x 的值已改变
[1] 2
```

第 4 章　绘图功能及基本统计

R 语言内建了许多绘图函数，这些函数可以显示各种统计图表，也可以用于自行绘制一些全新的图形。可以参考以下 R 语言的绘图示范：

```
> demo(graphics)
> demo(image)
```

R 语言的绘图指令可以分成以下三个基本类型。

（1）高级绘图（High-Level Plotting Functions）：绘制一个新的图形，包含坐标轴及标题等。

（2）低级绘图（Low-Level Plotting Functions）：用于在一个现有的图形上加上其他图形元素，如额外的点和线等。

（3）交互式绘图（Interactive Graphics Functions）：允许交互式地用其他设备（如鼠标）在一个现有的图形上加上图形信息。

本章重点内容：

- 高级绘图
- 交互式绘图
- 图形参数
- 基本统计

4.1　高级绘图

常用的高级绘图函数如表 4-1 所示。

表 4-1　常用的高级绘图函数

绘图函数	说明
plot(y)	以序号为横坐标（x 轴）、y 为纵坐标（y 轴）来绘图
plot(x,y)	以 x（x 轴）和 y（y 轴）来绘图
pie(y)	绘制饼形图
boxplot(y)	绘制盒形图
stem(y)	绘制茎叶图（Stem and Leaf Plot）

（续表）

绘图函数	说明
dotchart(y)	绘制点图
hist(y)	绘制直方图
barplot(y)	绘制条形图
contour(x, y, z)	绘制等高线图

高级绘图函数可以通过改变输入的自变量来产生不同的绘图效果。plot() 函数自变量的设置值如下：

```
plot(x, y,
type = "p",
bty="o",
pch =
lty =
cex =
lwd =
col =
bg =
xlim = NULL, ylim = NULL,
log = "",
main = NULL, sub = NULL,
xlab = NULL, ylab = NULL,
cex.main =
col.lab =
font.sub =
ann = par("ann"), axes = TRUE,
...)
```

其中：

- x　x 坐标位置。
- y　y 坐标位置。
- type　设置绘图在 (x,y)位置的显示方式。
  - type="p" 表示画点。
  - type="l" 表示画线。
  - type="b" 表示画点的同时在点与点之间画线连接。
  - type="s" 表示阶梯函数（Step Function），为左连续函数。type="S"表示阶梯函数，为右连续函数。
  - type="o" 表示画线的同时穿过画点。type="h"表示从点到 x 轴画垂直线。
  - type="n" 表示不画任何点与线，但容许画坐标轴来建立坐标系统，用于后面调用低级图形函数绘制图形。
- bty　表示设置图形坐标轴外框（Box）的类型，其选项有{ "o", "l", "7", "c", "u", "]" }。
- pch　表示画点时，设置 pch=k，k 是一个 1~25 的正整数，显示一个对应 k 的特定符号，其默认值为 1。
- lty　表示画线时，设置线条类型。

- > lty=1 表示实线（Solid Line）。
- > lty=2 表示虚线。
- lwd 表示画线时，设置线条宽度。
  - > lwd=1 表示默认值。
  - > lwd=k 表示线条宽度的倍数。
- col 表示设置点、线的颜色。默认值为 black，设置颜色值（Value）是调色板的数值或者颜色的名称。
- bg 表示设置图形的背景颜色，默认值为白色，例如 bg="white"。
- xlim 表示设置 x 坐标轴范围的上下界，如 xlim=c(1, 10)。
- ylim 表示设置 y 坐标轴范围的上下界，如 ylim=c(1, 10)。
- log 表示设置坐标轴是否取对数值。
  - > log="x" 表示对 x 坐标轴取对数值。
  - > log="y" 表示对 y 坐标轴取对数值。
  - > log="xy" 表示对 x 与 y 坐标轴都取对数值。
- main 表示设置图形主标题(Main Title)，定义的文字放在图形的上方，例如 main="Title"。
- sub 表示设置图形副标题（Subtitle），定义的文字放在图形的下方，例如 sub="Subtitle"。
- xlab 示设置 x 坐标轴标记，例如 xlab="X"。
- ylab 表示设置 y 坐标轴标记，例如 ylab="Y"。
- ann 表示是否画出自动设置的主标题及坐标轴标记，ann=TRUE 时会画出自动设置的主标题及坐标轴标记。
- axes 表示是否画出自动设置的坐标轴与坐标轴外框，axes=TRUE 时会画出自动设置的坐标轴与坐标轴外框。

例如：

```
> x <- sin(1:20)
> plot(x, type="l",main="Sin Plot", xlab="X",ylab="Y")
```

其结果如图 4-1 所示。

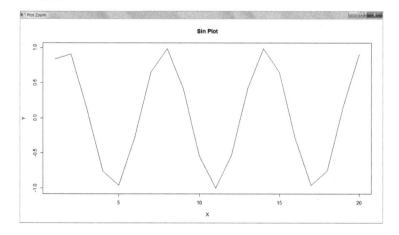

图 4-1 sin 图（正弦图）

4.2 低级绘图

常用的低级绘图函数如表 4-2 所示。

表 4-2 常用的低级绘图函数

绘图函数	说明
points(x,y)	在现有的图形上加一个点
lines(x,y)	在现有的图形上加一条线
text(x,y,labels=z.vec,...)	在(x,y)坐标点标出由 labels 设置的对应 z.vec 的数值型或文本型向量
abline(a,b)	在现有的图形上加画一条截距为 a 和斜率为 b 的直线
abline(h=y)	画出在 Y=y 平行于 x 坐标轴的水平线
abline(v=x)	画出在 X=x 垂直于 x 坐标轴的垂直线
polygon(x,y,...)	画出以(x,y)坐标点为顶点的多边形（Polygon），可以用 col=自变量指定一个特定颜色来填充多边形的内部
legend(x,y,leg.vec,...)	在现有的图形中指定(x,y)坐标位置绘制图例（Legend），图例的说明文字由向量 leg.vec 表示
title(main,sub)	main="My Main Title"设置图形主标题，定义的文字放在图形的上方，sub="My Subtitle"设置图形副标题，定义的文字放在图形的下方
mtext(text,side=3,line=1)	在现有图形的边缘加上文字

例如：

```
> x <- sin(1:20)
> plot(x, type="l", xlab="X",ylab="Y")
> title(main="Sin Plot",sub="图 4-2:低级绘图函数绘制的图 ")
```

其结果如图 4-2 所示。

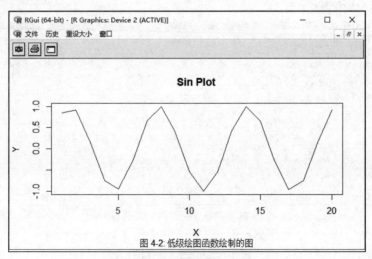

图 4-2 低级绘图函数绘制的图

4.3 交互式绘图

R 语言提供允许用户直接用鼠标在一个图形上提取（Extract）和添加（Add）信息的函数，其中最为常用的是 locator() 和 identify() 函数。

locator() 函数的功能是供用户单击当前图形上的特定位置：

```
locator(n,type)
```

其中：

- n 表示指定要单击几个坐标点，若不指定，则默认 n=512。
- type 允许在被选择的点上画图并且有高级绘图函数一样的效果，默认情况下不能画图。

locator() 函数使用列表对象包含元素 x 和 y 的方式返回所选中点的位置信息。

例如：

```
> plot(2, 2)
> pts <- locator(n = 3)    #用户可在图形中单击三个坐标点
> pts                      #单击完成后，显示出pts的值
$x
[1] 1.621502 1.840632 2.395573

$y
[1] 1.771951 2.347735 1.522875
```

identify() 函数的功能是容许用户将定义的标签（Label）使用鼠标左键放置在鼠标指针指向的位置。

```
identify(x,y,labels)
```

其中：

- x   x 坐标位置。
- y   y 坐标位置。
- labels   在未设置 labels 时，默认值为显示点的序号，选择完成后，右击结束选择的操作；当使用自变量 labels="My Labels" 时，可在单击处显示用户定义的标签"My Labels"，并右击结束鼠标选择的操作。

例如：

```
> x <- c(1, 3, 5, 7, 8, 9, 3, 6, 7, 2)
> y <- c(5, 3, 5, 8, 2, 1, 4, 3, 4, 7)
> plot(x, y)
> sel <- identify(x, y)  # 单击10次
```

其结果如图 4-3 所示。

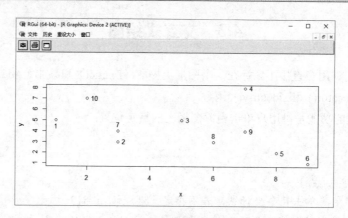

图 4-3 identify 图

在选择的过程中，identify()函数会在选择的数据点旁边标上用户定义的标签，而这些标签位置的序号会在选择完成之后返回给 sel 变量。最后，用户可以通过 sel 变量获取选择的数据点：

```
> x <- c(1, 3, 5, 7, 8, 9, 3, 6, 7, 2)
> y <- c(5, 3, 5, 8, 2, 1, 4, 3, 4, 7)
> plot(x, y)
> sel <- identify(x, y,"MY LABELS")    #单击选择图中间处
                                        #右击结束
> x.sel <- x[sel]
> y.sel <- y[sel]
> x.sel
[1] 5
> y.sel
[1] 5
```

其结果如图 4-4 所示。

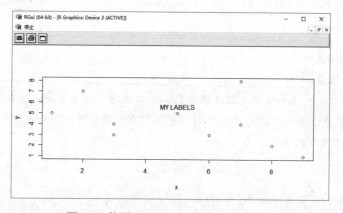

图 4-4 使用 identify(x,y,"MY LABELS")

4.4 图形参数

R 语言提供了许多图形参数（Graphics Parameters），用于控制图形的颜色、文字对齐等。直

接调用 par() 函数可以得到当前所有参数的设置值,例如:

```
> par()
```

通过 par() 函数可更改和设置图形参数:

```
par(par.name=par.value)
```

其中:

- par.name 表示图形参数的名称,可用于 par()或某些(高级或低级)绘图函数中作为自变量。
- par.value 表示设置给 par.name 图形参数的值。

例如:

```
> x <- c(5, 3, 5, 8, 2, 1, 4, 3, 4, 7)
> par(col=4, lty=4)
> plot(x, type="l", xlab="X",ylab="Y")
```

其结果如图 4-5 所示。

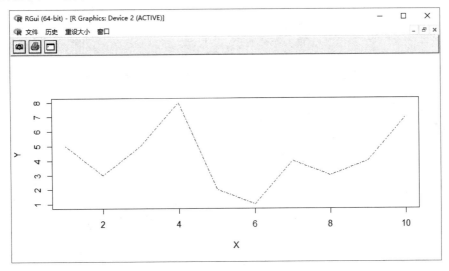

图 4-5 使用 par() 函数绘制的图

用户可使用 par(mar=c(bottom, left, top, right)) 设置图形离底部、左边、上边及右边的边界值,单位为英寸;也可使用 par(mfrow = c(nr, nc)) 表示显示 nr * nc 个子图。

例如:

```
par(mfrow=c(1,2))
par(mar=c(5, 4, 4, 2))
par(col=4, lty=1)

plot(x, type="l", xlab="X",ylab="Y")
barplot(x, xlab="X",ylab="Y")
```

其结果如图 4-6 所示。

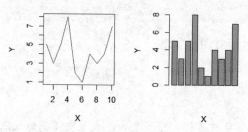

图 4-6　使用 par() 函数画 1*2 个子图

4.5　基本统计

在统计中，描述统计学（Descriptive Statistics）的主要目的就是通过数值或图形来呈现数据的特性并了解样本的统计特征。R 语言提供了 summary() 函数来获取数据及其分布的信息：

```
> summary(iris)
  Sepal.Length    Sepal.Width     Petal.Length    Petal.Width          Species  
 Min.   :4.300   Min.   :2.000   Min.   :1.000   Min.   :0.100   setosa    :50  
 1st Qu.:5.100   1st Qu.:2.800   1st Qu.:1.600   1st Qu.:0.300   versicolor:50  
 Median :5.800   Median :3.000   Median :4.350   Median :1.300   virginica :50  
 Mean   :5.843   Mean   :3.057   Mean   :3.758   Mean   :1.199                  
 3rd Qu.:6.400   3rd Qu.:3.300   3rd Qu.:5.100   3rd Qu.:1.800                  
 Max.   :7.900   Max.   :4.400   Max.   :6.900   Max.   :2.500                  
```

常用的统计图为直方图和盒形图。直方图又称柱状图，是用来呈现单个变量数据最常见的图，可有效展现数据的分布情况。

例如：

```
> y=c(170,170,171,172)
> hist(y,col='grey')
```

其结果如图 4-7 所示。

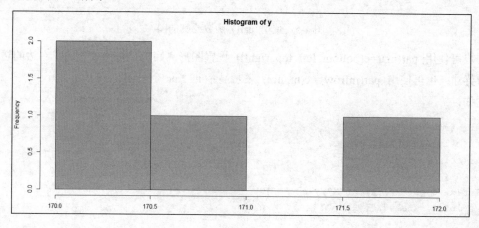

图 4-7　直方图

盒形图又称箱形图或盒须图,可显示出数据的最大值、最小值、中位数、第一个四分位数和第三个四分位数。若将数据从小到大排列并分成四等份,则会有三个分割点。从最小的那一边算起,第一个分割点称为第一个四分位数;第二个分割点称为第二个四分位数,也就是中位数;第三个分割点称为第三个四分位数。

例如:

```
> y1=c(165,166,167,167,175,176,177,178,179,180)
> median(y1,na.rm=TRUE)  # 中位数
[1] 175.5
```

其中,各项参数说明如下。

- na.rm=TRUE 表示若有 NA,则删除。

```
> max(y1)           # 最大值
[1] 180

> min(y1)           # 最小值
[1] 165

> max(y1)-min(y1)   # 全距
[1] 15

> range(y1)
[1] 165 180

> quantile(y1,0.25) 25%#   第一个四分位数
167

> quantile(y1,0.75)        # 第三个四分位数
75%
177.75

> IQR(y1)       # 四分位数间距
[1] 10.75
```

在垂直方向上画盒形图:

```
> boxplot(as.data.frame(y1), main = "boxplot(*, horizontal = FALSE)",
horizontal = FALSE)
```

其结果如图 4-8 所示。

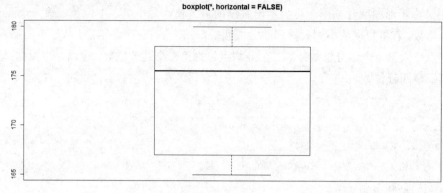

图4-8 盒形图

常用的描述性统计函数为均值（Mean）、中位数（Median）、众数（Mode）、方差（Variance）、标准差（Standard Deviation）及相关系数（Correlation）。

均值公式为：

$$\bar{x} = \frac{\sum_{i=1}^{N} x_i}{N} \qquad (4\text{-}1)$$

中位数为数据经过排序后的中间值。若样本数为偶数，则取中间两个数的平均值。众数是指一组数据中出现次数最多的那个数。

方差公式为：

$$var_x = \frac{\sum_{i=1}^{N}(x_i - \bar{x})^2}{N-1} \qquad (4\text{-}2)$$

标准差公式为：

$$S_x = \sqrt{var_x} \qquad (4\text{-}3)$$

相关系数公式为：

$$r_{xy} = \frac{\sum_{i=1}^{N}(x_i - \bar{x})(y_i - \bar{y})}{(N-1)S_x S_y} \qquad (4\text{-}4)$$

例如：

```
> y1=c(165,166,167,167,175,176,177,178,179,180)
> median(y1,na.rm = TRUE)  # 中位数
[1] 175.5

> var(y1)            # 方差
[1] 36

> sd(y1)             # 标准差
[1] 6

> table(y1)          # 出现次数
```

```
y1
165 166 167 175 176 177 178 179 180
  1   1   2   1   1   1   1   1   1

> which.max(table(y1))  # 众数及其排列位置
167
  3

> cor(y1,y1)        # 相关系数
[1] 1

> cor(y1,-y1)
[1] -1
```

回归（Regression）分析是一种统计分析方法，目的在于了解两个或多个变量间是否相关联，并建立数学模型，以观察特定变量来预测用户感兴趣的变量，其建立应变量（或称反应变量）与自变量之间关系的模型。线性回归是利用一个含有单个或多个自变量的回归公式来预测应变量，公式如下：

$$y=c_0+c_1\ x_1+c_2\ x_2+...+c_k\ x_k \tag{4-5}$$

其中：

- y　应变量（Dependent Variables）。
- c_i　回归系数（Regression Coefficients）。
- x_i　自变数（Independent Variables）。
- c_0　截距。

例如：

```
> setwd("D:/")       # 设置路径
> A10 <- read.table(file="grade.csv",header=TRUE,sep=",",encoding="GBK")
> str(A10)
'data.frame': 10 obs. of  3 variables:
$ 学生编号 : int  1 2 3 4 5 6 7 8 9 10
$ 每周自习时间 .X.: num  10.5 9.2 11.6 6.3 8.2 12.1 15.2 8.5 7.5 10.2
$ 考试成绩 .Y. : int  91 86 89 81 84 92 96 83 77 87

> A10 <- as.matrix(A10)   # 将 A10对象从数据框转成矩阵并删除 header
>  A10 <- matrix(A10, ncol = ncol(A10), dimnames = NULL)

> X=A10[,2]
> Y=A10[,3]
> Lm_model <- lm(Y ~ X)  # 执行回归
> Lm_model

Call:
lm(formula = Y ~ X)

Coefficients:
(Intercept)          X
     66.579      2.016
```

其回归公式可表示为：

$$Y = 66.579 + 2.016X \tag{4-6}$$

用户也可使用coef()来获取截距和回归系数，例如：

```
> cf <- coef(lm(Y ~ X))
> cf
(Intercept)           X
  66.579002    2.016213
```

若要验证回归公式的输出值，则可自定义函数或使用 sapply() 函数：

```
> lm_function <- function(x) {y <- cf[1]+cf[2]*x; return (y) }
> Y_output <- sapply(X,lm_function)
> Y_output
(Intercept) (Intercept) (Intercept) (Intercept) (Intercept) (Intercept) (Intercept)
   87.74924    85.12816    89.96708    79.28115    83.11195    90.97518    97.22544
(Intercept) (Intercept) (Intercept)
   83.71682    81.70060    87.14438
```

可使用 abs() 函数计算 Y-Y_output 的绝对值：

```
> abs(Y-Y_output)
(Intercept) (Intercept) (Intercept) (Intercept) (Intercept) (Intercept) (Intercept)
  3.2507584   0.8718357   0.9670761   1.7188541   0.8880489   1.0248172
(Intercept) (Intercept) (Intercept) (Intercept)
  1.2254439   0.7168150   4.7006018   0.1443776
```

使用 par() 函数设置画图参数，使用 plot() 和 abline() 函数画出回归公式，并使用points() 函数在图形上标示出 Y_output。

```
> par(mfrow=c(1,1))
> par(mar=c(5, 4, 4, 2))
> par(col="black")
> plot(X,Y)
> abline(lm(Y ~ X))
> par(col="blue")
> points(X,Y_output)
```

其结果如图4-9所示。

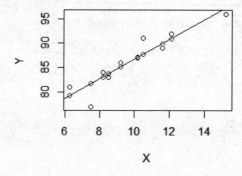

图4-9　回归公式图

第 5 章 相关程序包的介绍

本章介绍 R 语言在机器学习（Machine Learning）、数据挖掘（Data Mining）、文本挖掘（Text Mining）及大数据（Big Data）分析中常用的相关程序包。

本章重点内容：
- 机器学习
- 数据挖掘
- 社交网络分析及文本挖掘
- 大数据分析
- 程序包介绍

5.1 机器学习

顾名思义，机器学习就是让机器（计算机）具有学习能力，从数据中自动学会规则，并利用规则对新的数据进行预测。其主要用于设计和分析可以自动学习的算法，让计算机可以从过去的数据或经验中建立一个模型（Model），而学习（Learning）就是执行此模型，并使用训练数据集（Training Dataset）来建立模型。

常用的机器学习可以分成以下几种。

（1）监督式学习（Supervised Learning）：可以从训练数据中学到或建立一个模型，并按此模型预测新的案例。训练数据是由输入数据和预期输出所组成的。分类（Classification）就是常见的监督式学习算法。在机器学习领域中，可结合多个分类模型以达到更佳的分类性能，这种方法被称为组合方法（Ensemble Methods，或集成方法）。

（2）无监督式学习（Unsupervised Learning）：与监督式学习不同的是，训练数据中并无预期输出。聚类（Clustering）就是常见的无监督式学习算法。

（3）进化式学习（Evolutionary Learning）：主要是基于模仿生物进化及行为而发展出来的学习算法。基因算法（Genetic Algorithm）就是典型的进化式学习算法。

（4）混合式学习（Hybrid Learning）：主要是结合多种学习法的优点，以提升学习性能（Performance）或效率（Efficiency）。

5.2 数据挖掘

Fayyad 在 1996 年曾把数据库知识发现（Knowledge Discovery in Database, KDD）的过程定义为："从数据中建立确定有效的、新奇的、潜在有用的以及易懂形式的模型的过程，且此过程不是显而易见的"。数据挖掘为整个知识发现中最为核心的步骤，Berry & Linoff 在 1997 年将其定义为："数据挖掘是为了发现有意义的模型或规则（Rule），必须从大量数据中以自动或半自动的方式来探索和分析数据"。

常用的数据挖掘方法可以分成以下几种类别。

（1）分类（Classification）：将数据中各个属性（Attribute）分门别类地加以定义，通过训练大量数据后，用所得到的规则来建立类（Class）模型。分类属于监督式学习算法。

（2）聚类（Clustering）：通过相似程度的定义将数据分为不同的簇（Cluster）。其中相似程度可以利用不同的距离或相似度（Similarity）来定义。聚类与分类最大的不同在于：聚类并没有预先定义好类别，而聚类结果的意义要靠分析者事后的阐释才能决定。聚类属于无监督式学习算法。

（3）关联规则（Association Rule）：关联规则的目的是找出数据间可能相关的项目，通过数据寻找同时发生的事件（Event）或记录（Record）以推导出其间的关联规则。

R 语言中提供了许多与机器学习及数据挖掘有关的程序包。常用于分类的监督式学习算法包含决策树（Decision Tree）、支持向量机（Support Vector Machine）、人工神经网络（Artificial Neural Network）及组合方法（Ensemble Method），其对应的程序包为 rpart、C50、e1071、neuralnet、randomForest 和 adabag。常用于聚类的无监督式学习算法包含 K 均值聚类算法和模糊 C 均值聚类算法（Fuzzy C Means），对应的程序包为 e1071。进化式学习有许多不同的算法，其中包含基因算法及人工蜂群（Artificial Bee Colony）算法，对应的程序包为 GA 及 ABCoptim。常用的关联规则分析对应的程序包为 arules。

5.3 社交网络分析及文本挖掘

社群就是拥有相同兴趣或因共同目的而结合起来的团体。网络就是由一群特定功能的设备通过通信线路互相连接的节点达到资源共享的目的。社交网络的社群是一个建立在网络环境中的、虚拟的社会群体。

文本挖掘的特点在于原始输入数据都是没有特定结构的纯文本，这些文字的内容是由人类用自然语言写成的，无法直接使用数据挖掘的算法来探索和分析数据。

R 语言中提供了社交网络分析（Social Network Analysis）有关的程序包，可用于分析脸书（Facebook）数据，其对应的程序包为 Rfacebook 和 wordcloud。而文本挖掘的程序包则可使用 gutenbergr 和 jiebaR。

5.4 大数据分析

大数据一词最早由 IBM 公司提出，顾名思义，指的就是大量数据。大数据的 5 个特性分别是：大量（Volume）、高速（Velocity）、多样性（Variey）、低价值密度（Value）及真实性（Veracity）。大数据数据量大、速度快、多样性、价值密度低且数据的真伪难辨，所以需要全新的处理模式来促成更强的数据分析及决策能力。

Hadoop 是 Apache 软件基金会（Apache Software Foundation）的开放源码计划（Open source project），是以 Java 写成的分布式计算环境。Hadoop 软件平台可以将数据和运算的程序分散到可使用的不同计算机上，而且这些计算机的数量可以达到上千台之多。Hadoop 提供了 MapReduce 作为分布式处理技术，并提供了 HDFS（Hadoop Distributed File System）作为分布式文件技术，可以处理和存储大数据。用户可以在 http://hadoop.apache.org/中免费下载 Hadoop 软件。R 可与 Hadoop 紧密结合为 R+Hadoop，R+Hadoop 可使用 R 语言来使用 Hadoop 提供的技术。rmr2 是一个可以使用 R 语言在 Hadoop 上实现 MapReduce 的程序包。

Spark 也是 Apache 软件基金会（Apache Software Foundation）的开放源码计划（Open Source Project）。Spark 软件平台可以将数据和运算的程序分散到可使用的不同计算机上，而且这些计算机的数量可以达到上千台之多。用户可以在 http://spark.apache.org/中免费下载 Spark 软件。R 可与 Spark 紧密结合为 SparkR，SparkR 可通过 R 语言来使用 Spark 提供的技术。

5.5 程序包的介绍

本节将介绍常用的程序包：rpart、C50、e1071、neuralnet、randomForest、adabag、GA、ABCoptim、arules、Rfacebook、wordcloud、gutenbergr、jiebaR、rmr2、spark.kmeans、spark.mlp 及 spark.randomForest。

（1）rpart（Recursive partitioning for classification, regression and survival trees）程序包是由 Breiman、Friedman、Olshen and Stone 等人所撰写的 *Classification and Regression Trees* 书中所提出的重要方法。rpart 程序包中的 rpart() 函数的基本语法及其重要自变量如下：

```
rpart(formula, data, weights, subset, na.action=na.rpart, method, parms, control,...)
```

其中：

- formula　表示模型的公式，例如 $Y \sim X_1 + X_2 + ... + X_K$。
- data　表示为建立模型用的数据名称。
- weights　表示 data 的权重，为非必要性的自变量。
- subset　表示只使用部分子集合数据，为非必要性的自变量。
- na.action　表示对缺失值（NA）的处理方式，默认会删除。
- method　表示使用的方法，例如 anova、poisson、class 及 exp。
- parms　会根据上面不同的方法给予不同的自变量，若使用 anova 方法，则不需要设置此自变量。

- control 表示控制此算法的自变量，可在 rpart.control 中设置。

rpart.control() 函数的基本语法及其重要自变量如下：

```
rpart.control(minsplit=20, minbucket=round(minsplit/3), cp =0, maxdepth=30,...)
```

其中：

- minsplit 表示建立一个新节点（Node）时最少需要几项数据。
- minbucket 表示建立叶节点（Leaf Node）时最少需要几项数据。
- cp 决定计算复杂度（Complexity）的参数，用于修剪树的分支。
- maxdepth 表示树的深度。

rpart() 函数建立模型后，可调用 predict 函数来预测此模型的结果。predict() 函数的基本语法及其重要自变量如下：

```
predict(object, newdata, type, na.action,...)
```

其中：

- object 表示用来做预测的模型的名称。
- newdata 表示测试数据集的名称。
- type 表示使用的学习方法，例如 type="class" 表示使用分类方法。
- na.action 表示对缺失值的处理方式，默认不会删除（注意：与 rpart() 函数的处理方式不同）。

（2）C50 程序包实现了由 Quinlan 所提出的 C5.0 决策树。C50 程序包中的 C5.0() 函数的基本语法及其重要自变量如下：

```
C5.0(x, y, trials=1, rules=FALSE, weights=NULL, control= C5.0Control(),...)
```

其中：

- x 表示输入属性（Attribute）或应变量。
- y 表示输出属性或自变量（或解释变量）。
- trials 表示 boosting 的迭代（Iteration）次数，trials=1 表示只使用一种模型。
- rules 表示是否输出为规则而非树结构。
- weights 表示数据的权重。
- control 表示控制此算法的自变量，可在 C5.0Control() 中设置。

C5.0Control()函数的基本语法及其重要自变量如下：

```
C5.0Control(subset=TRUE,winnow=FALSE,noGlobalPruning=FALSE,CF,minCases,sample,
seed,label="outcome")
```

其中：

- subset 表示是否使用部分子集合数据。

- winnow 表示是否使用属性筛选。
- noGlobalPruning 表示是否进行决策树修剪。
- CF 表示置信水平，其值介于(0,1)之间。
- minCases 表示建立一个节点时最少需要几项数据（实例）。
- sample 表示当作训练数据的比例，其值介于 (0,0.999)之间。
- seed 随机数。
- label 表示输出属性的标签。

C5.0() 函数建立模型后，可调用 predict() 函数来预测此模型的结果。predict() 函数的基本语法及其重要自变量如下：

```
predict(object, newdata, trials, type, na.action,...)
```

其中：

- object 表示用来做预测的模型的名称。
- newdata 表示测试数据集的名称。
- trials 表示作为预测时，boosting 的迭代次数。
- type 使用 "class" 或 "prob"，例如 type="class" 表示使用分类。
- na.action 表示对缺失值的处理方式。

（3）e1071 程序包中包含支持向量机学习法。e10710 程序包中的 svm() 函数的基本语法及其重要自变量如下：

```
svm(formula, data, type, kernel, gamma, cost, subset, na.action, scale=TRUE,...)
```

其中：

- formula 表示模型的公式，例如 $Y \sim X_1+X_2+...+X_K$。
- data 表示建立模型用的数据的名称。
- type 表示使用的学习方法，包含 C-classification nu-classification、one-classification、eps-regression、nu-regression。
- kernel 表示核心（Kernel）函数，包含 linear、polynomial、radial base 及 sigmoid 函数。
- gamma 除了核心函数 linear 之外，其他核心函数使用的值。
- cost 表示 C 的值。
- cross 若 K > 0，则表示使用 K-fold。
- subset 表示使用部分子集合数据。
- na.action 表示对缺失值的处理方式。
- scale 表示属性数据是否要正规化。

svm() 函数建立模型后，可调用 predict() 函数来预测此模型的结果。predict() 函数的基本语法及其重要自变量如下：

```
predict.svm(object, newdata, na.action,...)
```

其中：

- object　表示用来做预测的模型的名称。
- newdata　表示测试数据集的名称。
- na.action　表示对缺失值的处理方式。

e1071 程序包中提供了 tune.svm() 函数搜索最优 gamma(γ) 和 cost(C) 值。tune.svm() 函数的基本语法及其重要自变量如下：

```
tune.svm(formula, data, gamma, cost, ...)
```

其中：

- formula　表示模型的公式。
- data　表示建立模型用的数据的名称。
- gamma　表示要搜索值的范围。
- cost　表示要搜索值的范围。

e1071 程序包中提供了执行模糊 C 均值聚类算法的 cmeans() 函数。cmeans() 函数的基本语法及其重要自变量如下：

```
cmeans(x, centers, iter.max, verbose, dist, method, m=2,...)
```

其中：

- x　表示作为聚类的数据。
- centers　表示簇的数量（Cluster Number）。
- iter.max　表示最大迭代次数。
- verbose　设为 TRUE 时，可显示执行过程的信息。
- dist　表示计算距离的公式，例如"euclidean" 和 "manhattan"。
- method　表示使用哪种学习方法，例如"cmeans"。
- m　为模糊 C 均值聚类算法的参数值。

（4）neuralnet 程序包提供了人工神经网络学习法。neuralnet 程序包中的 neuralnet() 函数的基本语法及其重要自变量如下：

```
neuralnet(formula, data, hidden, threshold, stepmax, rep, startweights, learningrate.
limit, learningrate, algorithm,...)
```

其中：

- formula　表示模型的公式。
- data　表示建立模型用的数据的名称。
- hidden　表示隐藏层神经元（Neuron）的数量（值得注意的是，此函数只提供了 1 层隐藏层）。
- threshold　表示临界值。
- stepmax　表示训练时最大迭代次数。

- rep 表示训练次数。
- startweights 表示开始时的权重值。
- learningrate.limit 表示学习率（Learning Rate）的最大和最小值。
- learningrate 表示当使用反向传播（Backpropagation）算法时的学习率。
- algorithm 表示使用的算法，例如 backprop、rprop+、rprop-、sag 或 slr。

neuralnet 程序包中提供了 compute() 函数来预测此模型的结果，compute() 函数的基本语法及其重要自变量如下：

```
compute(x, covariate, rep)
```

其中：

- x 表示用来做预测的模型的名称。
- covariate 表示测试数据集的名称。
- rep 表示预测次数。

（5）randomForest 程序包提供了组合方法的随机森林学习法。randomForest 程序包中的 randomForest() 函数的基本语法及其重要自变量如下：

```
randomForest(formula, data, ntree, na.action,...)
```

其中：

- formula 表示模型的公式。
- data 表示建立模型用的数据的名称。
- ntree 表示决策树的数量。
- na.action 表示数据中有缺失值时所使用的函数。

randomForest 程序包中提供了 predict() 函数来预测此模型的结果，predict() 函数的基本语法及其重要自变量如下：

```
predict(object, newdata,...)
```

其中：

- object 表示用来做预测的模型的名称。
- newdata 表示测试数据集的名称。

（6）adabag 程序包提供了组合方法的推进法（Boosting）。adabag 程序包中的 boosting() 函数的基本语法及其重要自变量如下：

```
boosting(formula, data, boos = TRUE, mfinal = 100,...)
```

其中：

- formula 表示模型的公式。
- data 表示建立模型用的数据的名称。

- boos 若设为 TRUE（默认值），则权重以该次迭代运算所使用训练集的观察值重新计算；若为 FALSE，则使用相同的权重。
- mfinal 表示迭代运算的次数。默认值为 mfinal=100（整数）。

adabag 程序包中提供了 predict() 函数来预测此模型的结果，predict() 函数的基本语法及其重要自变量如下：

```
predict(object, newdata, newmfinal,...)
```

其中：

- object 表示用来做预测的模型的名称。
- newdata 表示测试数据集的名称。
- newmfinal 表示用来做预测时的迭代运算次数。

NbClust 程序包中提供了 NbClust() 函数，用聚类指标（Clustering Index）来评估聚类的效果。NbClust() 函数的基本语法及其重要自变量如下：

```
NbClust (data, distance = "euclidean",min.nc=2, max.nc=15, method = "kmeans",index = "all",...)
```

其中：

- data 表示用于聚类的数据。
- distance 表示计算距离的公式，例如 euclidean 和 manhattan。
- min.nc 设置最小的簇数目（Cluster Number）。
- max.nc 设置最大的簇数目。
- method 表示使用哪种学习方法，例如 kmeans。
- Index 表示聚类指标，例如 kl、ch、hartigan、ccc、scott、marriot、trcovw、tracew、friedman、rubin、cindex、db、silhouette、duda、pseudot2、beale、ratkowsky、ball、ptbiserial、gap、frey、mcclain、gamma、gplus、tau、dunn、hubert、sdindex、dindex、sdbw，若为 all，则使用除了 gap、gamma、gplus 和 tau 之外的所有聚类指标。

（7）GA 程序包中提供了 ga() 函数执行基因算法。ga() 函数的基本语法及其重要自变量如下：

```
ga(type, fitness, min, max, nBits, population, selection, crossover, mutation, popSize, pcrossover, pmutation, elitism, monitor, maxiter, run, maxfitness,...)
```

其中：

- type 表示执行基因算法的方式，例如 binary、real-valued 和 permutation。
- fitness 表示适应度函数（Fitness Function）的值。
- min 表示搜索空间（Search Space）的最小值。
- max 表示搜索空间的最大值。
- nBits 表示 type="binary" 时的比特数。
- population 表示随机产生的初始族群（Population）。

- selection　表示基因算法的选择（Selection）机制。
- crossover　表示基因算法的组合交叉（Crossover）机制。
- mutation　表示基因算法的变异（Mutation）机制。
- popSize　表示族群的大小（Size）。
- pcrossover　表示产生组合交叉的概率。
- pmutation　表示发生变异的概率。
- elitism　表示多少百分比的个体（Individual）会保留到下一代（Generation）。
- monitor　显示执行的过程。
- maxiter　表示最大（迭）代次数。
- run　表示若连续几代都无法改善最优值，则停止执行基因算法。
- maxfitness　表示执行基因算法后找到的最大适应度函数值。

（8）ABCoptim 程序包中提供了 abc_optim() 函数执行人工蜂群算法。abc_optim() 函数的基本语法及其重要自变量如下：

```
abc_optim(par, fn, D, NP, FoodNumber,lb=-Inf, ub=+Inf, maxCycle=1000, criter=50,...)
```

其中：

- par　表示解的初始值。
- fn　表示求最小值的目标函数。
- D　表示解的参数（Parameter）数量。
- NP　表示人工蜂群的蜜蜂数量。
- FoodNumber　表示蜜蜂要搜索的食物数量。
- lb　表示搜索空间的下边界（Lower Bound）。
- ub　表示搜索空间的上边界（Upper Bound）。
- maxCycle　表示最大迭代次数。
- criter　表示结束条件。

（9）arules 程序包中提供了 apriori() 函数执行关联分析算法。apriori() 函数的基本语法及其重要自变量如下：

```
apriori(data, parameter = NULL, appearance = NULL, control = NULL)
```

其中：

- data　表示可以强制转换为事务历史记录的对象。
- parameter　表示参数，默认自变量值支持度（Support）= 0.1、置信度（confidence）= 0.8、最多规则数量 = 10。
- appearance　使用此自变量可以限制显示项目。默认情况下，所有项目都可以显示不受限制。
- control　表示控制此算法的自变量。

（10）使用 Rfacebook 程序包时必须先获得授权才能存取数据，Rfacebook 程序包中提供了 fbOAuth() 函数来产生一个可长期授权使用的 OAuth App ID 和 App Secret 来调用 Facebook（脸书）的 API。fbOAuth() 函数的基本语法及其重要自变量如下：

```
fbOAuth(app_id, app_secret, extended_permissions = TRUE)
```

其中：

- app_id　一组数字，表示由 OAuth 产生的应用程序编号，可到下列网址申请 https://developers.facebook.com/apps。
- app_secret　一组字符串，表示由 OAuth 产生的应用程序密钥。可到下列网址申请 https://developers.facebook.com/apps。
- extended_permissions　若设为 TRUE，则经授权过的用户可存取自己和好友的脸书私密（Private）信息，若设为 FALSE，则只能存取公开（Public）信息。

（11）使用 wordcloud 程序包可产生文字云，wordcloud()函数的基本语法及其重要自变量如下：

```
wordcloud(words,freq,min.freq=2,max.words=Inf,random.order=TRUE,...)
```

其中：

- words　要显示的文字。
- freq　要显示的文字出现的频率。
- min.freq　要显示的文字出现的最小频率。
- max.words　要显示的文字出现的最大频率。
- random.order　若设置为 TRUE，则显示文字的顺序会随机出现；若设置为 FALSE，则显示的文字会按频率递减出现。

（12）Project Gutenberg（PG）是把公版著作（Public Domain）数字化制作成电子书，放在网络上（https://www.gutenberg.org/）供用户自由取用。gutenbergr 程序包中可使用 gutenberg_download() 函数下载 Project Gutenberg 中数字化的文学作品或电子书：

```
gutenberg_download(gutenberg_id, strip,…)
```

其中：

- gutenberg_id　要下载的 Project Gutenberg ID，可为向量或数据框对象。
- strip　设为 TRUE 时，可去除 Header 和 Footer 内容。

（13）在进行文本挖掘时，中文与英文的处理方式有很大不同，英文有空格作为单词与单词之间自然分词的分隔符，而中文的词与词是连在一起的，所以在进行中文文本挖掘前，需要对文章中的句子进行分词处理，也就是将词与词分开以便进行进一步的分析。jiebaR 程序包中可调用 worker() 函数来初始化分词类型，调用 segment() 函数进行分词，基本语法及其重要自变量如下：

```
worker(type,dict,...)
```

其中：

- type　表示分词类型，其类型有 mix（混合）、mp（最大概率）、hmm、query（查询）、tag（标记）、simhash（敏感哈希）和 keywords（关键词）。默认的类型为混合模式。

- dict　表示主词典（Main Dictionary）的路径。

```
segment(code, jiebar,...)
```

其中：

- code　表示中文的文章或句子。
- jiebar　表示使用的 jiebaR Worker。

（14）rmr2 程序包中提供了 mapreduce() 函数和 rmr.options() 函数，让用户可使用 R 语言在 Hadoop 上实现 MapReduce。特别注意，使用此程序包时需先安装 Hadoop。Hadoop 的安装方式可参考附录 E。mapreduce() 和 rmr.options() 函数的基本语法及其重要自变量如下：

```
mapreduce(input,output,map,reduce,combine,input.format,output.format,verbose,...)
```

其中：

- input　表示输入数据。
- output　表示输出数据。
- map　表示使用的 map 函数。
- reduce　表示使用的 reduce 函数。
- combine　表示是否使用 combine 功能。
- input.format　表示输入数据格式。
- output.format　表示输出数据格式。
- verbose　设为 TRUE 时，可显示执行过程的信息。

```
rmr.options( backend = c("hadoop", "local"), profile.nodes, dfs.tempdir = NULL,
exclude.objects = NULL, backend.parameters = list())
```

其中：

- backend　表示选择以本地端（local）或 hadoop 模式执行程序。
- profile.nodes　表示搜集性能信息。
- dfs.tempdir　表示可存储中间结果。
- exclude.objects　表示可排除环境变量。
- backend.parameters　表示可直接传参数到 mapreduce。

（15）SparkR 中提供了 spark.kmeans()、spark.mlp() 和 spark.randomForest() 等函数（http://spark.apache.org/docs/latest/api/R/index.html），让用户可在 Spark 环境中使用 R 语言实现数据分析。特别注意，使用此程序包时需先安装 Spark 和 R for Linux，安装方式可参考附录 F。spark.kmeans()、spark.mlp() 和 spark.randomForest() 函数的基本语法及其重要自变量如下：

```
spark.kmeans(data,formula,k,maxIter,...)
```

其中：

- data　表示建立模型用的 SparkDataFrame 对象数据。

- formula 表示模型的公式，例如 $Y \sim X_1+X_2+...+X_K$。
- k 表示簇数目。
- maxIter 表示最大迭代次数。

```
spark.mlp(data,formula,layers,maxIter,...)
```

其中：

- data 表示建立模型用的 SparkDataFrame 对象数据。
- formula 表示模型的公式，例如 $Y \sim X_1+X_2+...+X_K$。
- layers 表示每一层的神经元数量。
- maxIter 表示最大迭代次数。

```
spark.randomForest(data,formula,type,maxDepth,numTrees,...)
```

其中：

- data 表示建立模型用的 SparkDataFrame 对象数据。
- formula 表示模型的公式，例如 $Y \sim X_1+X_2+...+X_K$。
- type 表示使用回归或分类决策树。
- maxDepth 表示决策树的最大深度。
- numTrees 表示决策树的数量。

第6章 监督式学习

本章介绍常用的监督式学习算法及其应用，例如决策树、支持向量机、人工神经网络及组合方法。

本章重点内容：
- 决策树
- 支持向量机
- 人工神经网络
- 组合方法

6.1 决策树

在机器学习的分类方法中，决策树可以说是最具代表性的方法。决策树具有不错的预测精度（Predict Accuracy）与可解释性（Interpretability），被各个领域广泛采用。决策树在建立过程中会建立一个树状的结构（简称树结构），其结构由根节点（Root Node）、子节点（Child Node）、叶节点（Leaf Node）或称为分类（Class）组成。决策树停止再往下衍生的情况分为该分类数据中的每一项数据都已经归类在同一分类下、该分类数据中已经没有办法再找到新的属性来进行节点分割或该分类数据中已经没有任何尚未处理的数据。当树结构建立后，新数据测试时是从决策树根部节点开始进行测试的，根据各分割选择属性值，移至另一子节点，按照此递归方式继续进行，直至到达叶节点，此叶节点就是一个预测的分类。从根节点由上而下经由子节点最后到达叶节点，即为一预测分类结果的规则。决策树的树结构图如图6-1所示。

如果原始训练数据属性太多或是决策树算法本身属性选择的偏好，就容易造成决策树过度拟合（Overfitting）的问题，以至于所产生的树结构太过于复杂，因此必须再进行适当的修剪（Pruning）。修剪的方式可分为预修剪（Pre-Pruning）和后修剪（Post-Pruning）。预修剪是运用统计方法来评估是否该继续分割某个子节点或立刻停止。后修剪允许决策树过度拟合的情况存在，建立决策树之后再来进行修剪。

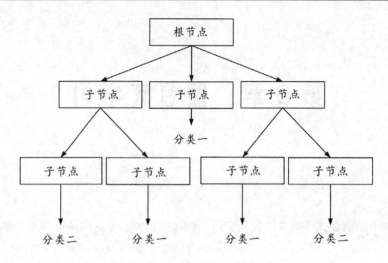

图 6-1　决策树的树结构图

目前，常见的决策树算法包括分类与回归树（Classification And Regression Trees，CART）、ID3（Inductive Dichotomiser 3）及 C5.0 等。

分类与回归树是 Breiman 等学者开发出的算法，其运算方式主要是利用二叉（Binary）分割原理。在分类与回归树运算过程中，永远分割出两个子节点，而这样的过程会被一再重复。因为每个分割出的子节点会继续分割出两个子节点，所以其算法以所有叶节点（分类）错误率的加权总和作为修剪的依据。

在 R 语言中，可调用 rpart 程序包中的 rpart() 函数来执行分类与回归树。程序范例 6-1 中调用 rpart() 函数来分类鸢尾花（Iris）数据，鸢尾花数据中共有 5 个属性，前 4 个属性为应变量，分别为花萼长度（Sepal.Length）、花萼宽度（Sepal.Width）、花瓣长度（Petal.Length）及花瓣宽度（Petal.Width）。第 5 个属性 Species 为应变量（目标属性），其代表鸢尾花的 3 个品种：Setosa、Virginica 和 Versicolor。

程序范例 6-1

首先使用 rpart 程序包和 iris 数据：

```
> library(rpart)
> data(iris)
```

使用 sample() 函数随机抽取 10% 的观察值（数据项数）用作测试数据：

```
> np = ceiling(0.1*nrow(iris))    # nrow(iris)返回iris数据项数
> np                               # ceiling()返回正向舍入的整数
[1] 15
> test.index = sample(1:nrow(iris),np)   # 10%为测试数据
> iris.testdata = iris[test.index,]      # 测试数据
> iris.traindata = iris[-test.index,]    # 训练数据
```

使用 rpart() 函数建立训练数据的决策树 iris.tree：

```
> iris.tree = rpart(Species ~ Sepal.Length + Sepal.Width +Petal.Length
+ Petal.Width, method="class", data=iris.traindata)
```

```
> iris.tree
n= 135
```

显示决策树 iris.tree 规则的信息（注意：范例中若随机抽取数据，则其执行结果可能不相同）：

```
>iris.tree

node), split, n, loss, yval, (yprob)
      * denotes terminal node

1) root 135 88 virginica (0.3259259 0.3259259 0.3481481)
2) Petal.Length< 2.45 44  0 setosa (1.0000000 0.0000000 0.0000000) *
3) Petal.Length>=2.45 91 44 virginica (0.0000000 0.4835165 0.5164835)
6) Petal.Width< 1.75 49  5 versicolor (0.0000000 0.8979592 0.1020408) *
7) Petal.Width>=1.75 42  0 virginica (0.0000000 0.0000000 1.0000000) *
```

显示决策树 iris.tree 的 cp 值、错误率及各节点的详细信息：

```
> summary(iris.tree) Call:
rpart(formula = Species ~ Sepal.Length + Sepal.Width + Petal.Length + Petal.Width, data =
iris.traindata, method = "class")
  n= 135

      CP nsplit          rel error    xerror      xstd
1 0.5056180      0 1.00000000 1.1910112 0.05361608
2 0.4382022      1 0.49438202 0.6629213 0.06475539
3 0.0100000      2 0.05617978 0.1235955 0.03571497

Variable importance
Petal.Width Petal.Length Sepal.Length  Sepal.Width
         34            31           21           14

Node number 1: 135 observations,    complexity param=0.505618
  predicted class=setosa      expected loss=0.6592593  P(node) =1
    class counts:    46     45     44
   probabilities: 0.341 0.333 0.326
  left son=2 (46 obs) right son=3 (89 obs)
  Primary splits:
      Petal.Length < 2.45 to the left,  improve=45.49080, (0 missing)
      Petal.Width  < 0.8  to the left,  improve=45.49080, (0 missing)
      Sepal.Length < 5.45 to the left,  improve=31.08528, (0 missing)
      Sepal.Width  < 3.05 to the right, improve=17.89879, (0 missing)
   Surrogate splits:

      Petal.Width  < 0.8  to the left,  agree=1.000, adj=1.000, (0 split)
      Sepal.Length < 5.45 to the left,  agree=0.919, adj=0.761, (0 split)
      Sepal.Width  < 3.35 to the right, agree=0.830, adj=0.500, (0 split)

Node number 2: 46 observations
  predicted class=setosa       expected loss=0  P(node) =0.3407407
    class counts:  46     0     0
```

```
        probabilities: 1.000 0.000 0.000

Node number 3: 89 observations,    complexity param=0.4382022
    predicted class=versicolor  expected loss=0.494382  P(node) =0.6592593
      class counts:    0    45    44
     probabilities: 0.000 0.506 0.494
    left son=6 (48 obs) right son=7 (41 obs)
    Primary splits:
        Petal.Width  < 1.75 to the left,  improve=35.209830, (0 missing)
        Petal.Length < 4.75 to the left,  improve=33.934380, (0 missing)
        Sepal.Length < 6.15 to the left,  improve=10.447780, (0 missing)
        Sepal.Width  < 2.95 to the left,  improve= 3.518872, (0 missing)
    Surrogate splits:
        Petal.Length < 4.75 to the left,  agree=0.899, adj=0.780, (0 split)
        Sepal.Length < 6.15 to the left,  agree=0.730, adj=0.415, (0 split)
        Sepal.Width  < 2.95 to the left,  agree=0.674, adj=0.293, (0 split)

Node number 6: 48 observations
    predicted class=versicolor  expected loss=0.08333333  P(node) =0.3555556
      class counts:    0    44     4
     probabilities: 0.000 0.917 0.083

Node number 7: 41 observations
    predicted class=virginica   expected loss=0.02439024  P(node) =0.3037037
      class counts:    0     1    40
     probabilities: 0.000 0.024 0.976
```

画出决策树并标示文字，其结果如图 6-2 所示。

```
> plot(iris.tree); text(iris.tree)
```

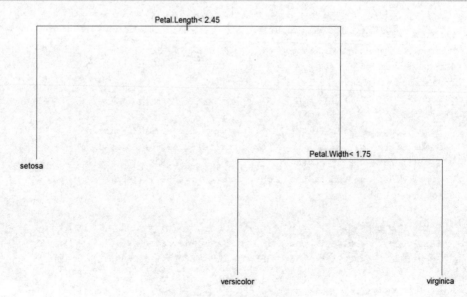

图 6-2　Iris 的树结构图

显示训练数据的正确率：

```
> species.traindata = iris$Species[-test.index]
> train.predict=factor(predict(iris.tree, iris.traindata,
+ type='class'), levels=levels(species.traindata))
>
> table.traindata =table(species.traindata,train.predict)
> table.traindata
                 train.predict
species.traindata setosa versicolor virginica
       setosa         47          0         0
       versicolor      0         42         1
       virginica       0          5        40
```

从以上数据得知：47 项 setosa 都分类正确，有 1 项 versicolor 被错分到 virginica，有 5 项 virginica 被错分到 versicolor，所以正确率为（47+42+40）/135 * 100% = 95.55556%。

```
> correct.traindata=sum(diag(table.traindata))/sum(table.traindata)*100
#diag() 返回对角线元素值，sum() 返回加总值
> correct.traindata
[1] 95.55556  #训练数据正确率 95.55556%
```

显示测试数据的正确率：

```
> species.testdata = iris$Species[test.index]
> test.predict=factor(predict(iris.tree, iris.testdata,
+ type='class'), levels=levels(species.testdata))
> table.testdata =table(species.testdata,test.predict)
> table.testdata
                test.predict
species.testdata setosa versicolor virginica
       setosa        7          0         0
       versicolor    0          3         0
       virginica     0          1         4
>
```

从以上数据得知：测试数据共有 15 项（观察值），setosa 和 versicolor 都分类正确，有 1 项 virginica 被错分到 versicolor，所以正确率为（7+3+4）/15 * 100% = 93.33333%。

```
> correct.testdata=sum(diag(table.testdata))/sum(table.testdata)*100
#diag()返回对角线元素，sum()返回加总值
> correct.testdata
[1] 93.33333          #测试数据正确率 93.33333%
```

程序范例 6-2

用户可设置合适的 rpart.control() 函数的自变量来改善决策树的分类结果，例如 rpart.control(minsplit=5, cp=0.0001, maxdepth=30)。

```
> library(rpart)
> data(iris)
>
```

```
> np = ceiling(0.1*nrow(iris))    # 10% 为测试数据
> np
[1] 15
>
> test.index = sample(1:nrow(iris),np)
>
> iris.testdata = iris[test.index,]    # 测试数据
> iris.traindata = iris[-test.index,]  # 训练数据
>
> iris.tree = rpart(Species ~ Sepal.Length + Sepal.Width +Petal.Length
+ Petal.Width, method="class",  data=iris.traindata,
+ control=rpart.control(minsplit=5, cp=0.0001, maxdepth=30) )
>
> species.traindata = iris$Species[-test.index]
> train.predict=factor(predict(iris.tree, iris.traindata,
+ type='class'), levels=levels(species.traindata))
>
> table.traindata =table(species.traindata,train.predict)
> table.traindata
                  train.predict
species.traindata setosa versicolor virginica
       setosa         47          0         0
       versicolor      0         44         1
       virginica       0          2        41
> correct.traindata=sum(diag(table.traindata))/sum(table.traindata)*100
> correct.traindata
[1] 97.77778    #训练数据正确率为 97.77778%
>
> species.testdata = iris$Species[test.index]
> test.predict=factor(predict(iris.tree, iris.testdata,
+ type='class'), levels=levels(species.testdata))
> table.testdata =table(species.testdata,test.predict)
> table.testdata
                 test.predict
species.testdata setosa versicolor virginica
       setosa         3          0         0
       versicolor     0          5         0
       virginica      0          0         7
> correct.testdata=sum(diag(table.testdata))/sum(table.testdata)*100
> correct.testdata
[1] 100                          #测试数据正确率为 100%
>
> iris.tree
n= 135

node), split, n, loss, yval, (yprob)
      * denotes terminal node

1) root 135 88 setosa (0.34814815 0.33333333 0.31851852)
2) Petal.Length< 2.45 47  0 setosa (1.00000000 0.00000000 0.00000000) *
```

```
3) Petal.Length>=2.45 88 43 versicolor (0.00000000 0.51136364 0.48863636)
6) Petal.Width< 1.75 49  5 versicolor (0.00000000 0.89795918 0.10204082)
12) Petal.Length< 4.95 43  1 versicolor (0.00000000 0.97674419 0.02325581) *
13) Petal.Length>=4.95 6  2 virginica (0.00000000 0.33333333 0.66666667)
26) Petal.Width>=1.55 3  1 versicolor (0.00000000 0.66666667 0.33333333) *
27) Petal.Width< 1.55 3  0 virginica (0.00000000 0.00000000 1.00000000) *
7) Petal.Width>=1.75 39  1 virginica (0.00000000 0.02564103 0.97435897) *
```

ID3 使用信息增益（Information Gain）当作属性选择指标（Selection Measure），而 C5.0 是 Quinlan 学者依据 ID3 算法改进而来的。C5.0 和 ID3 一样是以信息增益最大的属性为分割属性，两者最大的差异在于 ID3 偏向选择属性值较多的属性，而 C5.0 使用增益比例（Gain Ratio）当作属性选择指标。信息增益计算方式如下：

$$Info(S) = -\sum_{i=1}^{k}\left\{\left[freq(C_i,S/|S|)\right]\log_2\left[freq(C_i,S/|S|)\right]\right\} \tag{6-1}$$

$$Info_x(S) = -\sum_{i=1}^{L}\left[(|S_i|/|S|)Info(S_i)\right] \tag{6-2}$$

$$Gain(X) = Info(S) - Info_x(S) \tag{6-3}$$

$$SplitInfo(X) = -\sum_{i=1}^{L}\left[\frac{|S_i|}{|S|}\log_2\frac{|S_i|}{|S|}\right] \tag{6-4}$$

$$GainRatio(X) = Gain(X)/SplitInfo(X) \tag{6-5}$$

在公式(6-1)中，S 为在训练数据中各分类的总个数，C 为目标属性中所含有的分类，S_i 为分类下包含的项数；在公式(6-2)中，S 为在选定属性下训练数据中各分类的总个数，L 为输入的属性数量，$Info_x(S)$代表在某一情况选择下一个候选属性的熵值（Entropy）；在公式（6-3）中，Gain(X)代表信息增益，Info(S) 为未增加某一属性下的熵值。

C5.0 初始时利用公式 (6-1) 计算无考虑属性下的信息混乱度；公式 (6-2) 为考虑某一属性情况下信息的混乱度；公式 (6-3) 为公式 (6-1) 与公式 (6-2) 相减后信息的混乱度之差，即信息增益；公式 (6-4) 为在某一情况下分割节点的熵值；公式 (6-5) 为其增益比例。

在建树的过程中，决策树算法利用最小案例数量（Minimum Case，MC）检验每个节点所包含的案例数量是否超过 M，若没超过，则该节点停止生长；反之继续往下成长。当树结构生长完后，为避免造成过度拟合问题，同时降低树的复杂度，必须进行树的修剪。C5.0 采用后修剪法，在修剪阶段利用设置修剪（Pruning）的置信水平（Confidence Level，CF）下的置信区间作为基准，计算该子节点以及下一个节点统计上预期错误率与案例数量的置信水平，当目前子节点错误率小于下一个子节点时，统计显示该节点继续分割下去也不会有更好的结果，便将其修剪掉。

程序范例 6-3

首先使用 C50 程序包和 iris 数据：

```
> library(C50)
```

```
> data(iris)
> np = ceiling(0.1*nrow(iris))          # 10% 为测试数据
> np
[1] 15
```

随机读取 10% 为测试数据，90% 为训练数据：

```
> test.index = sample(1:nrow(iris),np)
> iris.test = iris[test.index,]         # 测试数据
> iris.train = iris[-test.index,]       # 训练数据
```

设置 C5.0 相关的自变量：

```
> c=C5.0Control(subset = FALSE,
+               bands = 0,
+               winnow = FALSE,
+               noGlobalPruning = FALSE,
+               CF = 0.25,
+               minCases = 2,
+               fuzzyThreshold = FALSE,
+               sample = 0,
+               seed = sample.int(4096, size = 1) - 1L,
+               earlyStopping = TRUE
+              )
>                           #第 5 个属性为目标属性
> iris_treeModel <- C5.0(x = iris.train[, -5], y = iris.train$Species, control.=c)
```

显示 C5.0 决策树 iris.treeModel 的规则和训练数据的错误率：

```
> summary(iris_treeModel)
Call:
C5.0.default(x = iris.train[, -5], y = iris.train$Species, control = c)


C5.0 [Release 2.07 GPL Edition]
-------------------------------

Class specified by attribute `outcome'

Read 135 cases (5 attributes) from undefined.data
Decision tree:
Petal.Length <= 1.7: setosa (41)
Petal.Length > 1.7:
:...Petal.Width > 1.7: virginica (43/1)
    Petal.Width <= 1.7:
    :...Petal.Length <= 5.3: versicolor (49/1)
        Petal.Length > 5.3: virginica (2)

Evaluation on training data (135 cases):

        Decision Tree
```

```
                -----------------
                 Size      Errors

                    4      2( 1.5%)    <<


                  (a)   (b)   (c)    <-classified as
                  ----  ----  ----
                   41                   (a): class setosa
                         48     1       (b): class versicolor
                          1    44       (c): class virginica
            Attribute usage:

             100.00% Petal.Length
              69.63% Petal.Width


Time: 0.0 secs
```

由以上得知，训练数据的错误率为 1.5%（正确率为 98.5%）：

```
>test.output=predict(iris_treeModel, iris.test[, -5], type = "class")
>n=length(test.output)
>number=0
>for( i in 1:n)           #计算测试数据的正确率
+{
+if (test.output[i] == iris.test[i,5])
+    {
+    number=number+1
+    }
+}
>test.accuracy=number/n*100
>test.accuracy
[1] 73.33333
```

由以上得知，测试数据的正确率为 73.33333%。

需特别注意的是：C5.0 输出变量（例如 iris.train$Species）的数据类型必须是因子，若是非因子数据类型，则可使用 factor() 函数来转换。

```
> iris.train$Species=factor(iris.train$Species)
```

程序范例 6-4

用户也可直接设置 C5.0Control()函数的 sample 自变量来代表训练数据的项数，例如 sample = 0.9 表示 90% 为训练数据。

```
> library(C50)
> library(stringr)
>
> data(iris)
>
> c=C5.0Control(subset = FALSE,
```

```
+                       bands = 0,
+                       winnow = FALSE,
+                       noGlobalPruning = FALSE,
+                       CF = 0.25,
+                       minCases = 2,
+                       fuzzyThreshold = FALSE,
+                       sample = 0.9,       # 90%为训练数据
+                       seed = sample.int(4096, size = 1) - 1L,
+                       earlyStopping = TRUE,
+                       label = "Species")
> iris_treeModel <- C5.0(x = iris[, -5], y = iris$Species,
+                  control =c)
>
> summary(iris_treeModel)

Call:
C5.0.default(x = iris[, -5], y = iris$Species, control = c)


C5.0 [Release 2.07 GPL Edition]
-------------------------------


Class specified by attribute 'Species'

Read 135 cases (5 attributes) from undefined.data
Decision tree:
Petal.Length <= 1.9: setosa (45)
Petal.Length > 1.9:
:...Petal.Width > 1.7: virginica (41/1)
    Petal.Width <= 1.7:
    :...Petal.Length <= 4.9: versicolor (43/1)
        Petal.Length > 4.9: virginica (6/2)
Evaluation on training data (135 cases):

            Decision Tree
          ----------------
          Size     Errors
            4     4( 3.0%)   <<


          (a)    (b)    (c)    <-classified as
         ----   ----   ----
          45                   (a): class setosa
                 42     3      (b): class versicolor
                  1    44      (c): class virginica


     Attribute usage:

     100.00% Petal.Length
      66.67% Petal.Width
```

```
Evaluation on test data (15 cases):

            Decision Tree
          ----------------
          Size      Errors

            4        0( 0.0%)   <<

          (a)    (b)    (c)    <-classified as
          ----   ---    ----
            5                  (a): class setosa
                   5            (b): class versicolor
                          5    (c): class virginica

Time: 0.0 secs
```

调用 stringr 程序包中的 str_locate_all() 函数和 substr() 函数可得到测试数据的错误率。具体方法是先调用 str_locate_all() 函数获取列表 iris_treeModel 的 output 元素(iris_treeModel$output)中 << 的位置，再由 su+bstr() 函数获取 % 前一个位置到前四个位置的文字（表示测试数据的错误率），最后调用 as.numeric() 函数转变为数字。

```
> tt=as.character(iris_treeModel$output)    #转为文字
> x=str_locate_all(tt,"<<")
> y=substr(tt,x[[1]][2]-9,x[[1]][2]-6)

> test.error=as.numeric(y)
> test.correct=100-test.error
> test.correct       #测试数据正确率
[1] 100
```

6.2 支持向量机

支持向量机（Support Vector Machine，SVM）是由 Vapnik 于 1995 年和 AT&T 实验室团队在提出的统计学习理论（Statistical Learning Theory）中发展出的学习算法。支持向量机的学习架构是以小样本的训练数据来得到最优的学习与归纳能力，使得 SVM 能够学习出较为平滑的曲线，使得之后的测试数据也能在最少的变化中进行分类（Classification）或回归（Regression），即结构风险最小化（Structure Risk Minimization，SRM）原则。

以图 6-3 所示的二维空间为例，其中白点与黑点分别代表两类训练数据。对于 H_1、H_2、H_3 三条线，每一条线都可视为一个分类器。用户可找出 H_2 分类器，使得这两类产生最大的距离以得到最小的分类错误率，故 H_2 为此图中最优的分类器。在高维度空间中，SVM 可寻找一个超平面（Hyper-Plane）分类器，以得到最小的分类错误率，SVM 在解决非线性分类的问题方面也有极佳的处理能力。

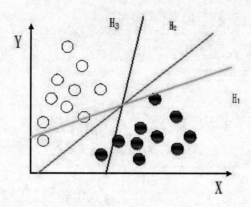

图 6-3 分类示意图

对于分类的问题，是寻找一个函数 $y=f(x)$，$x \in R_n$，$y\in \{1,-1\}$，在给定的 k 个训练数据 $(x_1,y_1)\sim(x_i,y_i)$、$i=1\sim k$ 中，寻找一个最优超平面，可将训练数据加以分类，计算公式如下：

$$(w \cdot x) + b = 0, w \in R_n, b \in R \tag{6-6}$$

此平面满足下列条件：

$$y_i[w \cdot x_i + b] > 0, i = 1 \sim k \tag{6-7}$$

此时为了避免因为信息噪声而使得误差过大，必须在此加入松弛变量 $\zeta_i \geqslant 0$，使得计算公式变成：

$$y_i[w \cdot x_i + b] > 1 - \xi_i, i = 1 \sim k \tag{6-8}$$

而此时被超平面分割成两部分的训练数据，其各自到达超平面的最小距离为最大，计算方法如下：

$$Min: [\gamma(w) = \frac{1}{2} w \cdot w + C \sum_{i=1}^{k} \xi_i] \tag{6-9}$$

而此有限制的优化问题可以通过 Lagrange 乘数（Multiplier）转化为对偶问题：

$$L = \frac{1}{2}\|w\|^2 + C\sum_{i=1}^{k}\xi_i - P - \sum_{i=1}^{k}\gamma_i \xi_i \tag{6-10}$$

$$P = \sum_{i=1}^{k} \alpha\{y_i(w \cdot x_i + b) - 1 + \xi_i\} \tag{6-11}$$

其中，L 包含 4 个参数（w、b、α、γ），需满足对 w、b 最小化且对 α、γ 最大化，而经由 Karush-Kuhn-Tucker（KKT）理论即可求得 $\sum_{i=1}^{k} \alpha_i y_i = 0$、$w = \sum_{i=1}^{k} \alpha_i y_i x_i$，由此可得到最大化函数：

$$Max: \left[\sum_{i=1}^{k} \alpha_i - \frac{1}{2} \sum_{i,j=1}^{k} \alpha_i \alpha_j y_i y_j (x_i \cdot x_j) \right] \tag{6-12}$$

限制条件为:

$$0 \leq a_i \leq C, i=1\sim k, \sum_{i=1}^{k} a_i y_i = 0$$

SVM 可将数据从 $k(u,v) = (\Phi(x_i), \Phi(x_j))$ 转换到高维度的特征空间中,所以 SVM 处理优化的问题可以转变为:

$$L_D = \sum_{i=1}^{m} \alpha_i - \frac{1}{2} \sum_{i,j=1}^{m} \alpha_i \alpha_j y_i y_j (\Phi(x_i), \Phi(x_j)) \tag{6-13}$$

$$L_D = \sum_{i} \alpha_i - \frac{1}{2} \sum_{i,j} \alpha_i \alpha_j y_i y_j k(u,v) \tag{6-14}$$

限制条件为:

$$0 \leq a_i \leq C, i=1\sim k, \sum_{i=1}^{k} a_i y_i = 0$$

其中,$k(u,v)$ 为核心函数,常用的核心函数包含 linear、polynomial、radial base 和 sigmoid 函数,linear 函数定义为 $k(u,v) = u'v$,polynomial 函数定义为 $k(u,v) = (u'v + coef0)^{degree}$,radial basis 函数定义为 $k(u,v) = e^{(-\gamma|u-v|^2)}$,sigmoid 函数定义为 $k(u,v) = \tanh(ru'v + coef0)$。

程序范例 6-5

首先使用 e1071 程序包和 iris 数据:

```
> library(e1071)
> data(iris)
> index <- 1:nrow(iris)
> np = ceiling(0.1*nrow(iris))    # 10% 为测试数据
> np
[1] 15
```

随机读取 10% 为测试数据及 90% 为训练数据:

```
> testindex = sample(1:nrow(iris),np)
> testset = iris[test.index,]        # 测试数据
> trainset = iris[-test.index,]      # 训练数据
```

使用训练数据建立 svm.model 模型,核心函数为 radial base 函数,并设置自变量 cost(C)=10、gamma(γ)=10:

```
> svm.model <- svm(Species ~ ., data = trainset, type = 'C-classification', cost = 10, gamma = 10)
```

使用 svm.model 模型来预测测试数据:

```
> svm.pred <- predict(svm.model, testset[,-5])
```

显示测试数据的正确率:

```
> table.svm.test=table(pred = svm.pred, true = testset[,5])
```

```
> table.svm.test
              true
pred         setosa  versicolor  virginica
  setosa        2         0           0
  versicolor    0         7           0
  virginica     1         1           4
> correct.svm=sum(diag(table.svm.test))/sum(table.svm.test)
> correct.svm=correct.svm*100
> correct.svm
[1] 86.66667    #测试数据正确率
```

e1071 程序包中提供了 tune.svm() 函数，可搜索最优的 cost(C) 和 gamma(γ) 值。用户可设置搜索范围，例如 $0.001 \leq \gamma \leq 0.1$ 和 $0.1 \leq C \leq 10$：

```
> tuned <- tune.svm(Species ~., data = trainset, gamma = 10^(-3:-1), cost = 10^(-1:1))
> summary(tuned)

Parameter tuning of 'svm':

- sampling method: 10-fold cross validation

- best parameters:
 gamma cost
   0.1   10

- best performance: 0.04450549

- Detailed performance results:
   gamma    cost    error       dispersion
1  0.001    0.1     0.74395604  0.13837721
2  0.010    0.1     0.42362637  0.22152013
3  0.100    0.1     0.14065934  0.08211993
4  0.001    1.0     0.41648352  0.22753121
5  0.010    1.0     0.13351648  0.07800779
6  0.100    1.0     0.05879121  0.05843897
7  0.001   10.0     0.13351648  0.07800779
8  0.010   10.0     0.05219780  0.06125272
9  0.100   10.0     0.04450549  0.06261755
```

由以上结果得知：$\gamma = 0.1$ 和 $C = 10$ 时误差值最小，所以选定此值作为 svm() 函数的参数并重新计算正确率。

```
> model <- svm(Species ~., data = trainset, kernel="radial", gamma=0.1, cost=10)
> summary(model)

Call:
svm(formula = Species ~ ., data = trainset, kernel = "radial",
    gamma = 0.1, cost = 10)


Parameters:
   SVM-Type:  C-classification  SVM-Kernel:  radial
       cost:  10
      gamma:  0.1

Number of Support Vectors:  31
```

```
( 3 15 13 )
Number of Classes:  3

Levels:
 setosa versicolor virginica

>
> svm.pred <- predict(model, testset[,-5])
>
> ## compute svm confusion matrix
> table.svm.best.test=table(pred = svm.pred, true = testset[,5])
> table.svm.best.test
           true
pred        setosa     versicolor     virginica
  setosa         3              0             0
  versicolor     0              8             0
  virginica      0              0             4
> correct.svm.best=sum(diag(table.svm.best.test))/sum(table.svm.best.test)*100
> correct.svm.best
[1] 100   #测试数据正确率 100%
```

6.3 人工神经网络

人工神经网络是由许多人工神经细胞与其连接所组成的，并且可以组成各种网络模型（Network Model），而人工神经细胞又称为人工神经元、处理单元。处理单元（Processing Element，PE）是人工神经网络最基本的组成单位，每一个处理单元的输出必须连接到下一层的处理单元。处理单元的输入、输出值的计算公式可用下列函数来表示，公式表示如下：

$$f_i = f(net_i) = f(\sum_i W_{ij} X_i - \theta_i) \tag{6-15}$$

其中：

- f_i 人工神经网络处理单元的输出信号。
- net_i 集成函数。
- f 人工神经网络处理单元的转换函数。
- W_{ij} 人工神经网络各处理单元间的连接权重值（Weight）。
- X_i 输入向量。
- θ_i 人工神经网络处理单元的阈值（Threshold）。

人工神经网络的架构中被广泛应用的是反向传播（Back Propagation，BP）人工神经网络，这是一种监督式学习的人工神经网络，基本原理是利用最陡坡降法（Gradient Steepest Descent Method）以迭代方式将误差函数最小化。反向传播算法的网络训练包含两个阶段：正向传播阶段和反向传播阶段，如图 6-4 所示。在正向传播阶段，输入向量从输入层开始输入，并以正向传播方式经由隐藏层传至输出层，最后计算出推论输出值。在此阶段，网络节点之间的键值都是固定的。在反向传播阶段，网络节点之间的连接权重值则根据误差修正规则来进行修正。通过连接权重值的修正，使修正后的推论输出值能偏向于目标输出值。误差函数的公式表示如下：

$$E = \frac{1}{2}\sum_j (d_j - y_j)^2 \qquad (6\text{-}16)$$

其中：

- d_j　输出层第 j 个输出单元的目标输出值。
- y_j　输出层第 j 个输出单元的推论输出值。

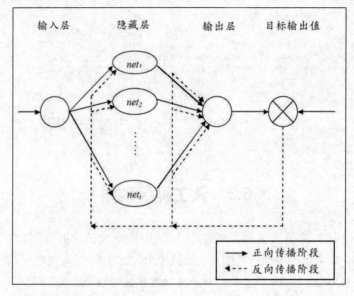

图6-4　反向传播算法的正向传播阶段和反向传播阶段

误差函数对权重值偏微分的公式表示如下：

$$\Delta W = -\eta \frac{\partial E}{\partial W} \qquad (6\text{-}17)$$

其中：

- ΔW　代表各层处理单元间的连接权重值的修正量。
- η　称为学习速率（Learning Rate），主要用来控制每次权重值修改的大小。

误差函数对隐藏层第 k 个单元与输出层第 j 个单元间的连接权重值关系可用如下公式表示：

$$\Delta W_{kj} = \eta \times \delta_j \times X_k \qquad (6\text{-}18)$$

其中：

- X_k　第 k 个隐藏层单元输入向量。
- δ_j　区域梯度函数，$\delta_j = (d_j - y_j) y_j (1 - y_j)$。

误差函数对输入层第 i 个单元与隐藏层第 k 个单元间的连接权重值关系可用如下公式表示：

$$\Delta W_{ik} = \eta \times \delta_k \times X_i \tag{6-19}$$

其中：

- δk　区域梯度函数 $\delta_k = y(1-y_k) \sum_i \delta_i W_{ik}$。

程序范例 6-6

首先使用 neuralnet 程序包：

```
> library("neuralnet")
```

使用均匀分布（Uniform Distribution）产生 100 个 0~100 的训练数据，并计算其平方根值作为目标值（输出值）：

```
> Var1 <- runif(100, min=0, max=100)
> sqrt.data <- data.frame(Var1, Sqrt=sqrt(Var1))
```

设置 neuralnet() 函数的隐藏层处理单元的自变量 hidden = 10、阈值 = 0.01、学习速率 = 0.01 以及使用反向传播人工神经网络，并将执行结果赋值给 net.sqrt：

```
> net.sqrt <- neuralnet(Sqrt~Var1, sqrt.data, hidden=10,threshold=0.01)
```

打印 net.sqrt 相关信息：

```
> print(net.sqrt)
Call: neuralnet(formula = Sqrt ~ Var1, data = sqrt.data, hidden = 10, threshold = 0.01)

1 repetition was calculated.

      Error Reached Threshold Steps
1 0.0001658715951  0.009936046689 10537
```

画出 net.sqrt 的架构图，如图 6-5 所示。

```
> plot(net.sqrt)
```

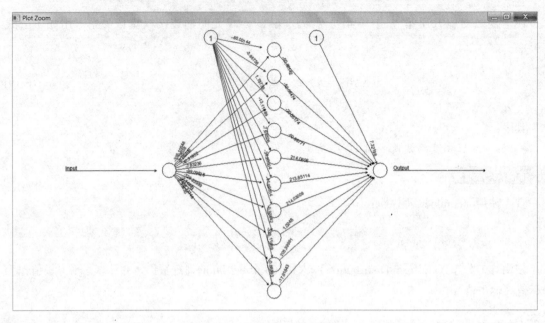

图 6-5 net.sqrt 的架构图

产生测试数据并使用 compute() 函数来预测此模型结果：

```
> testdata <- as.data.frame((1:10)^2)
> nn.result <- compute(net.sqrt, testdata)
```

打印预测结果：

```
> print(nn.result)
$neurons
$neurons[[1]]
      1 (1:10)^2
 [1,] 1        1
 [2,] 1        4
 [3,] 1        9
 [4,] 1       16
 [5,] 1       25
 [6,] 1       36
 [7,] 1       49
 [8,] 1       64
 [9,] 1       81
[10,] 1      100
$neurons[[2]]
     [,1]                                                     [,2]         [,3]         [,4]
[1,] 1 0.257725812205483739703026913048233836889266967734 0.9919478656 0.6044046623
[2,] 1 0.022929844482689979573430960613222850952297449111 9 0.9907018379 0.8125646725
[3,] 1 0.000263126189463150274988612054016812180634588003 2 0.9881873662 0.9610199115
[4,] 1 0.000000489781480088627602662368354380362234223866 8 0.9835047641 0.9964540598
[5,] 1 0.000000000151233491949292229408925714029265918725 3 0.9747324334 0.9998442540
[6,] 1 0.000000000000007748445630382309519748607051781164 0 0.9577294311 0.9999965980
[7,] 1 0.000000000000000000006587223541027999251275260039 76 0.9235472752 0.9999999629
```

```
 [8,] 1 0.00000000000000000000000000009292047371886474438178180 0.8539429308 0.9999999998
 [9,] 1 0.00000000000000000000000000000002174911704725879390 0.7197913330 1.0000000000
[10,]     1 0.0000000000000000000000000000000000000084468162670 0.5060569487 1.0000000000
```

```
                      [,5]                 [,6]            [,7]                  [,8]
 [1,] 0.802116690204665583 0.7968909277964 0.2510113653 0.557289991528997842849
 [2,] 0.675954358580697079 0.7449406210565 0.2677665348 0.328123030451510755243
 [3,] 0.408061586830988920 0.6410360878618 0.2972179318 0.091556009044004516006510
 [4,] 0.127627524893795391 0.4728123320481 0.3414099410 0.010942303497950861546077
 [5,] 0.019548653245394374 0.2700522148165 0.4024439148 0.000645597421561323425550
 [6,] 0.001741956148366336 0.1113886025399 0.4811573388 0.000020065286517870950710
 [7,] 0.000098064227900182 0.0337093301416 0.5750879525 0.000000331529709919523830
 [8,] 0.000003539693612293 0.0079112815363 0.6767441203 0.000000002913719960693670
 [9,] 0.000000082042026592 0.0014949996732 0.7743813690 0.000000000013621650034020
[10,] 0.000000001221126965 0.0002308477486 0.8564052781 0.000000000000033874143350
```

```
                [,9]          [,10]         [,11]
 [1,] 0.541182501700 0.313699456493 0.3014415717
 [2,] 0.504031573736 0.200055996536 0.3211166332
 [3,] 0.442217067074 0.248221939978 0.3553398072
 [4,] 0.358982541606 0.198976939161 0.4057762028
 [5,] 0.263721441826 0.146966628131 0.4735095293
 [6,] 0.171794037122 0.099231314426 0.5573786856
 [7,] 0.098098407455 0.060979131897 0.6521084818
 [8,] 0.049106005126 0.034088025562 0.7478745784
 [9,] 0.021718638784 0.017373953135 0.8330633738
[10,] 0.008567711444 0.008100501284 0.8992465457
```

```
$net.result
             [,1]
 [1,] 1.029557458
 [2,] 2.000101230
 [3,] 2.998497421
 [4,] 4.001433416
 [5,] 4.998696195
 [6,] 6.001244074
 [7,] 6.999690285
 [8,] 7.998276198
 [9,] 9.002924905
[10,] 9.990079622
```

若要计算平均绝对误差（mae）和均方根误差（rmse），则可再载入 DMwR 程序包并调用 regr.eval() 函数。

```
> library(DMwR)

> regr.eval(expected.output,nn.result$net.result[,1],
+ stats=c('mae','rmse'))

        Mae          rmse
0.05002136169 0.03148412327
```

6.4 组合方法

Nilsson 在 1965 年提出：由多位专家组合而成，按一些特定方式（如投票法、权重法）整合各位专家的意见进行决策，其得到的结果会比只有单个专家的效果更好。由于每位专家的擅长之处不同，因此通过组合的机制可以让专家之间彼此互补，得到更好的结果。常用的组合方法包含套袋法（Bagging）和推进法（Boosting）。

6.4.1 随机森林

套袋法是将所有预测模型的多数预测值当作未知值分组的预测值，主要是以放回重复（With Replacement）随机选取方式从原始数据集合中选取与原集合相同数量的数据产生多个训练集合，使用投票法并按多数投票来决定分类结果。

随机森林（Random Forest）是一个特别设计给决策树分类法使用的套袋法，结合多个决策树的预测结果，而每棵树都是根据随机森林的随机向量的值所建立的。

程序范例 6-7

首先清除 R 软件内存中的数据（确认没有上次使用过的对象）：

```
> rm(list = ls())
> gc()
```

使用 randomForest 程序包和 iris 数据集：

```
> library(randomForest)
> data(iris)
```

使用 sample() 函数将 iris 数据分为训练数据集（80%）和测试数据集（20%）：

```
> ind <- sample(2, nrow(iris), replace=TRUE, prob=c(0.8, 0.2))
> trainData <- iris[ind==1,]
> testData <- iris[ind==2,]
```

使用训练集数据和 100 棵决策树建立随机森林：

```
> rf <- randomForest(Species ~ ., data=trainData, ntree=100)
```

使用 predict() 函数来预测测试数据的正确率：

```
> irisPred <- predict(rf, newdata=testData)
> table(irisPred, testData$Species)

irisPred     setosa  versicolor  virginica
  setosa       13        0           0
  versicolor    0        9           0
  virginica     0        0           9
```

由以上数据得知，测试数据的正确率为 100%。

6.4.2 推进法

推进法是对每个训练数据集都设置一个权重，每次迭代后，对分类错误的数据加大权重，使得下一次的迭代更加关注这些数据，推进法采用不放回（Without Replacement）的随机抽取法。

程序范例 6-8

使用 adabag 程序包和 iris 数据集：

```
> library(adabag)
> data(iris)
```

使用 sample()函数将 iris 数据分为训练数据集（80%）和测试数据集（20%）：

```
> ind <- sample(2, nrow(iris), replace=TRUE, prob=c(0.8, 0.2))
> trainData <- iris[ind==1,]
> testData <- iris[ind==2,]
```

使用训练集数据以及迭代 5 次来建立推进法：

```
> train.adaboost <- boosting(Species~., data=trainData, boos=TRUE, mfinal=5)
```

调用 predict() 函数来预测测试数据的正确率：

```
> test.adaboost.pred <- predict.boosting(train.adaboost,newdata=testData)
> test.adaboost.pred$confusion
               Observed Class
Predicted Class setosa versicolor virginica
     setosa         7        0          0
     versicolor     0        9          0
     virginica      0        1          9
> test.adaboost.pred$error
[1] 0.03846154
```

由以上数据得知，测试数据的正确率为 25/26*100%= 96.15%。

第 7 章 无监督式学习

本章介绍常用的无监督式学习算法及其应用,例如层次聚类法(Hierarchical Clustering)、K 平均聚类算法(K Means)及模糊 C 平均聚类算法(Fuzzy C Means)。

本章重点内容:

- 层次聚类
- K 平均算法
- 模糊 C 平均算法
- 聚类指标

7.1 层次聚类法

聚类(Clustering)是将一组数据根据相似度计算公式将其聚合或分割成若干群。层次聚类法是通过如图 7-1 所示的层次架构方式将数据层层反复地进行聚合或分割,以产生最后的树结构。常见的方式有两种:聚合法(Agglomerative)采用由下而上的处理方式,从树结构的底部开始,将数据或者各个簇逐次合并,一开始将每个数据都视为一个独立的簇,然后根据簇间相似度计算公式不断地合并两个最相似的簇,直到最后所有的簇都合并成一个大的聚类为止;而分割法(Divisive)采用的是自上而下的处理方式,从树结构的顶端开始将大的簇逐次分割,一开始将所有数据视为一个大的簇,同样不断地根据相似度计算公式将大簇分割成较小的簇,直到簇数目达到事先所设定的数目为止。

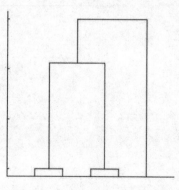

图 7-1 层次聚类法的树结构

层次聚类法的相似度可由"距离"来计算,两个簇 (c_i, c_j) 相似度(Similarity)的计算方式如下:

$$\text{Similarity} = \frac{1}{1+d(c_i, c_j)} \tag{7-1}$$

其中两簇之间的距离 $d(c_i, c_j)$ 可使用图 7-2 提供的 4 种方法来计算：最短距离法（Single Linkage），即获取簇和簇之间最近点的距离 $d_{min}(c_i, c_j)$；重心法（Centroid Linkage），即获取簇和簇之间中心点的距离 $d_{mean}(c_i, c_j)$；平均法（Average Linkage），即获取簇和簇之间所有点距离的平均距离 $d_{avg}(c_i, c_j)$；最长距离法（Complete Linkage），即获取簇和簇之间最远点的距离 $d_{max}(c_i, c_j)$。对于点和点之间的距离，则可使用欧几里得距离（Euclidean Distance）、皮尔森相系数（Pearson Correlation Coefficient）或马氏距离（Mahalanobis Distance）。

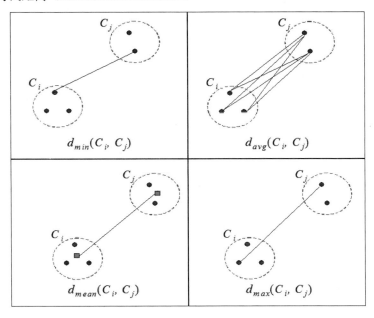

图 7-2 计算各聚类之间的距离

程序范例 7-1

为了方便展示层次聚类法的树结构，本范例只随机取用 10%（15）项 iris 数据当作样本：

```
> data(iris)
> index <- 1:nrow(iris)
> np <- ceiling(0.1*nrow(iris))    # 使用 10%数据当样本
> idx <- sample(1:nrow(iris),np)
> irisSample <- iris[idx,]
```

由于层次聚类法是无监督式学习法，因此将 irisSample 目标属性 Species 设为 NULL：

```
> irisSample$Species <- NULL
```

使用 hclust() 函数并设置使用单个连接法来计算簇距离：

```
> hc <- hclust(dist(irisSample), method="single")
```

使用 plot() 函数绘出树结构并标示目标属性 Species 的品种，其结果如图 7-3 所示。

```
> plot(hc,labels=iris$Species[idx])
```

使用 rect.hclust() 函数将 3 个簇以矩形标示出来：

```
> rect.hclust(hc, k=3)
```

最后的结果如图 7-4 所示。

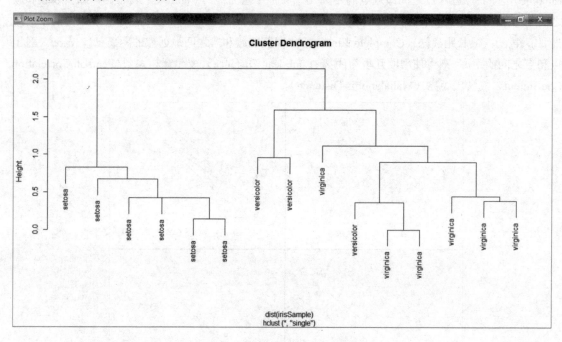

图 7-3 层次聚类法的 iris 树结构

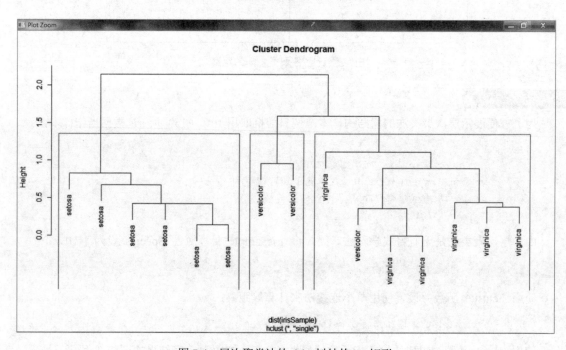

图 7-4 层次聚类法的 iris 树结构 + 矩形

7.2 K 平均聚类算法

K 平均聚类算法是 MacQueen 于 1967 年提出的聚类算法,必须事前设定簇的数量为 k,k 个簇 C_i, $i=1,2,\ldots,k$,计算公式 (7-2) 用来达到聚类的优化目的。

$$argmin \sum_{i=1}^{k} \sum_{x_j \in c_i} \|x_j - \mu_i\| \tag{7-2}$$

其中:

- μ_i 第 i 个簇的质心。

K 平均聚类算法按照下列 4 个步骤进行聚类:

（1）给定 k 值,将数据分割成 k 个非空子集合。
（2）计算现在分割簇的质心,这些质心为各个簇的中心点。
（3）将每个数据归类于最接近的质心点。
（4）回到步骤 2,一直到每个簇数据没有任何变化。

程序范例 7-2

首先加载 iris 数据并显示其属性:

```
> data(iris)
> attributes(iris)
$names
[1] "Sepal.Length" "Sepal.Width"  "Petal.Length" "Petal.Width"  "Species"
$row.names
 [1]   1   2   3   4   5   6   7   8   9  10  11  12  13  14  15  16  17  18
 19  20  21  22  23  24  25  26  27  28  29
 [30]  30  31  32  33  34  35  36  37  38  39  40  41  42  43  44  45  46  47
 48  49  50  51  52  53  54  55  56  57  58
 [59]  59  60  61  62  63  64  65  66  67  68  69  70  71  72  73  74  75  76
 77  78  79  80  81  82  83  84  85  86  87
 [88]  88  89  90  91  92  93  94  95  96  97  98  99 100 101 102 103 104 105
 106 107 108 109 110 111 112 113 114 115 116
[117] 117 118 119 120 121 122 123 124 125 126 127 128 129 130 131 132 133 134
 135 136 137 138 139 140 141 142 143 144 145
[146] 146 147 148 149 150

$class
[1] "data.frame"
```

将 iris 赋值给 iris2 并将其目标属性设为 NULL:

```
> iris2 <- iris
> iris2$Species <- NULL
```

调用 kmeans() 函数将 iris2 数据分为 3 个簇并直接显示结果(注意:使用小括号括住表达式):

```
> (kmeans.result <- kmeans(iris2, 3))
K-means clustering with 3 clusters of sizes 50, 38, 62
```

```
Cluster means:
  Sepal.Length Sepal.Width Petal.Length Petal.Width
1     5.006000    3.428000     1.462000    0.246000
2     6.850000    3.073684     5.742105    2.071053
3     5.901613    2.748387     4.393548    1.433871
Clustering vector:
  [1] 1 1 1 1 1 1 1 1 1 1 1 1 1 1 1 1 1 1 1 1 1 1 1 1 1 1 1 1 1
 [30] 1 1 1 1 1 1 1 1 1 1 1 1 1 1 1 1 1 1 1 1 1 3 3 2 3 3 3 3 3
 [59] 3 3 3 3 3 3 3 3 3 3 3 3 3 3 3 3 3 2 3 3 3 3 3 3 3 3 3 3 3
 [88] 3 3 3 3 3 3 3 3 3 3 3 2 3 2 2 2 2 2 3 2 2 2 2 2 2 3 3 2
[117] 2 2 2 3 2 3 2 2 2 3 2 2 2 2 3 2 2 2 2 3 2 2 2 3 2 2 2 3 2 2
[146] 2 3 2 2 3

Within cluster sum of squares by cluster:
[1] 15.15100 23.87947 39.82097
 (between_SS / total_SS =  88.4 %)

Available components:

[1] "cluster"      "centers"     "totss"      "withinss"
[5] "tot.withinss" "betweenss"   "size"       "iter"
[9] "ifault"
```

调用 table() 函数显示聚类的结果,从以下结果得知有 36 个 versicolor 聚类到 virginica,有 48 个 virginica 聚类到 versicolor:

```
> table(iris$Species, kmeans.result$cluster)
              1   2   3
  setosa     50   0   0
  versicolor  0   2  48
  virginica   0  36  14
```

以属性 Sepal.Length 为 X 轴、Sepal.Width 为 Y 轴,使用 plot() 函数和 points() 函数画出簇的质心位置,如图 7-5 所示。

```
> plot(iris2[c("Sepal.Length", "Sepal.Width")], col = kmeans.result$cluster)
> points(kmeans.result$centers[,c("Sepal.Length", "Sepal.Width")], col = 1:3,
pch = 8, cex=2)
```

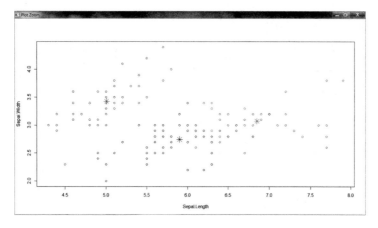

图 7-5 质心位置图

7.3 模糊 C 平均聚类算法

Bezdek 于 1973 年提出模糊 C 平均聚类算法，其目标函数应用了模糊理论的概念，使得每一个输入向量不再仅归属于某一特定的簇，而以其归属程度来表现隶属于各个簇的程度。

数据集 $X = \{x_1, x_2, ..., x_n\}$，模糊 C 平均聚类算法将数据集分成 c 簇。令这个 c 簇的质心集为 $V = \{v_1, v_2, ..., v_c\}$。令数据点 x_i 对簇质心 v_i 的隶属函数值为 u_{ij}，隶属矩阵 U 表示为：

$$U = \begin{bmatrix} u_{11} & u_{12} & \cdots & u_{1n} \\ u_{21} & u_{22} & \cdots & u_{2n} \\ \vdots & \vdots & \vdots & \vdots \\ u_{c1} & u_{c2} & \cdots & u_{cn} \end{bmatrix} \tag{7-3}$$

质心集 V 和数据点集 X 的误差为：

$$E(U, V : X) = \sum_{i=1}^{c} \sum_{j=1}^{n} (u_{ij})^m \|x_j - v_i\|^2 \tag{7-4}$$

限制条件为：

$$\sum_{i=1}^{c} u_{ij} = 1 \tag{7-5}$$

根据 Lagrange 乘数方法，可得：

$$L(U, \lambda) = \sum_{j=1}^{n} \sum_{i=1}^{c} (u_{ij})^m \|x_j - v_i\|^2 - \sum_{j=1}^{n} \lambda_j \left(\sum_{i=1}^{c} u_{ij} - 1 \right) \tag{7-6}$$

$L(U, \lambda)$ 对 λ_j 微分后，令其为零，可得到：

$$\frac{\partial L(U, \lambda)}{\partial \lambda_j} = 0 \Leftrightarrow \sum_{i=1}^{c} u_{ij} - 1 = 0 \tag{7-7}$$

$L(U, \lambda)$ 对 U_{ij} 微分后，令其为零，可得到：

$$\frac{\partial L(U,\lambda)}{\partial u_{ij}} = 0 \Leftrightarrow \left[m(u_{ij})^{m-1}\|x_j - v_i\|^2 - \lambda_j\right] = 0 \tag{7-8}$$

由公式 (7-8) 可解得：

$$u_{ij} = \left(\frac{\lambda_j}{m\|x_j - v_i\|^2}\right)^{\frac{1}{m-1}} \tag{7-9}$$

由公式 (7-7) 和公式 (7-9) 可得到：

$$\sum_{i=1}^{c} u_{ij} = \sum_{i=1}^{c}\left(\frac{\lambda_j}{m\|x_j - v_i\|^2}\right)^{\frac{1}{m-1}} = 1 \tag{7-10}$$

从公式 (7-10) 可得：

$$\left(\frac{\lambda_j}{m}\right)^{\frac{1}{m-1}} = 1 \bigg/ \sum_{i=1}^{c}\left(\frac{1}{m\|x_j - v_i\|^2}\right)^{\frac{1}{m-1}} \tag{7-11}$$

将公式 (7-11) 代入公式 (7-9)，得到：

$$u_{ij} = 1 \bigg/ \sum_{k=1}^{c}\left(\frac{\|x_j - v_i\|}{\|x_j - v_k\|}\right)^{\frac{2}{m-1}} \tag{7-12}$$

质心 v_i 可调整为：

$$v_i = \frac{\sum_{j=1}^{n}(u_{ij})^m x_j - 1}{\sum_{j=1}^{n}(u_{ij})^m}, 1 \leq i \leq c \tag{7-13}$$

模糊 C 平均聚类算法按照下列 4 个步骤进行聚类：

（1）设置簇数 c、次方 m、误差容忍度 ε 和初始隶属矩阵 U_0。
（2）根据数据集和 U_0 算出初始的质心集。
（3）重新计算 u_{ij}，$1 \leq i \leq c$ 和 $1 \leq j \leq n$，修正各个质心值。
（4）计算出误差 $E = \sum_{i=1}^{c}\|v_i^{t+1} - v_i^t\|$，若小于 ε，则停止；否则回到步骤 3。

程序范例 7-3

首先使用 e1071 程序包和 iris 数据：

```
> library("e1071")
> data(iris)
```

调用 rbind() 函数将 iris 数据赋值给 x 并调用 t() 函数来进行矩阵的转置：

```
> x<-rbind(iris$Sepal.Length, iris$Sepal.Width, iris$Petal.Length, iris$Petal.Width)
> x<-t(x)
```

执行 cmeans() 函数并设置自变量聚类的簇数 centers=3、m=2、最大迭代次数 iter.max=5 及 verbose=TRUE 来显示执行过程的信息：

```
> result<-cmeans(x,m=2,centers=3,iter.max=500,verbose=TRUE,method="cmeans")
Iteration:    1, Error:   1.1065549797
Iteration:    2, Error:   0.5143839622
Iteration:    3, Error:   0.4317889832
Iteration:    4, Error:   0.4089850507
Iteration:    5, Error:   0.4047123528
Iteration:    6, Error:   0.4038350861
Iteration:    7, Error:   0.4035582273
Iteration:    8, Error:   0.4034493445
Iteration:    9, Error:   0.4034040642
Iteration:   10, Error:   0.4033850717
Iteration:   11, Error:   0.4033771134
Iteration:   12, Error:   0.4033737857
Iteration:   13, Error:   0.4033723967
Iteration:   14, Error:   0.4033718175
Iteration:   15, Error:   0.4033715763
Iteration:   16, Error:   0.4033714758
Iteration:   17, Error:   0.4033714340
Iteration:   18, Error:   0.4033714166
Iteration:   19, Error:   0.4033714094
Iteration:   20 converged, Error:  0.4033714063
> print(result)
Fuzzy c-means clustering with 3 clusters

Cluster centers:
       [,1]     [,2]     [,3]     [,4]
1  5.003966 3.414092 1.482811 0.2535443
2  5.888876 2.761049 4.363869 1.3972723
3  6.774943 3.052362 5.646696 2.0535137

Memberships:
              1           2           3
 [1,] 0.996623591 0.0023043863 0.0010720223
 [2,] 0.975851052 0.0166506280 0.0074983195
```

```
 [3,] 0.979824951 0.0137602302 0.0064148190
 [4,] 0.967425471 0.0224665169 0.0101080122
 [5,] 0.994470320 0.0037617495 0.0017679303
 [6,] 0.934570922 0.0448085703 0.0206205077
 [7,] 0.979490658 0.0140045709 0.0065047707
 [8,] 0.999547236 0.0003115527 0.0001412117
 [9,] 0.930376026 0.0477206324 0.0219033420
[10,] 0.982721838 0.0119363182 0.0053418439
[11,] 0.968040923 0.0217577753 0.0102013018
[12,] 0.992136600 0.0054321973 0.0024312023
[13,] 0.970638460 0.0201839493 0.0091775909
[14,] 0.922966474 0.0517966675 0.0252368585
[15,] 0.889755281 0.0726119584 0.0376327603
[16,] 0.841339798 0.1043526293 0.0543075725
[17,] 0.946923379 0.0355805082 0.0174961132
[18,] 0.996652682 0.0022885250 0.0010587927
[19,] 0.904130899 0.0655599590 0.0303091423
[20,] 0.979188234 0.0141579180 0.0066538478
[21,] 0.968602441 0.0218567895 0.0095407691
[22,] 0.984832257 0.0103737118 0.0047940313
[23,] 0.958656627 0.0275101878 0.0138331853
[24,] 0.979425664 0.0144624401 0.0061118961
[25,] 0.966916808 0.0232444617 0.0098387303
[26,] 0.973566137 0.0184599651 0.0079738984
[27,] 0.994844951 0.0035816529 0.0015733961
[28,] 0.993347956 0.0045652424 0.0020868017
[29,] 0.993675739 0.0043282354 0.0019960261
[30,] 0.979514890 0.0142022292 0.0062828813
[31,] 0.978726304 0.0147999074 0.0064737890
[32,] 0.974363071 0.0176971498 0.0079397789
[33,] 0.938518560 0.0411467611 0.0203346789
[34,] 0.904167145 0.0634837012 0.0323491536
[35,] 0.985064315 0.0103414539 0.0045942311
[36,] 0.984993039 0.0102016329 0.0048053285
[37,] 0.964183139 0.0242524395 0.0115644217
[38,] 0.990890771 0.0061826377 0.0029265911
[39,] 0.939681223 0.0410855538 0.0192332228
[40,] 0.998288945 0.0011777899 0.0005332648
[41,] 0.994727723 0.0035839504 0.0016883265
[42,] 0.850728207 0.1022757303 0.0469960626
[43,] 0.952613146 0.0321779226 0.0152089315
[44,] 0.979286701 0.0143835533 0.0063297452
[45,] 0.945268152 0.0381305220 0.0166013258
[46,] 0.972144726 0.0192234127 0.0086318609
[47,] 0.976792090 0.0158427458 0.0073651640
[48,] 0.974219574 0.0176533601 0.0081270657
[49,] 0.977219457 0.0155207770 0.0072597662
[50,] 0.997072388 0.0020082807 0.0009193315
[51,] 0.044575050 0.4542004553 0.5012244951
[52,] 0.029165723 0.7639574496 0.2068768275
```

```
[53,]  0.031265512 0.3686821493 0.6000523388
[54,]  0.049339121 0.8702294370 0.0804314421
[55,]  0.024110917 0.7586613416 0.2172277416
[56,]  0.005739272 0.9737967869 0.0204639407
[57,]  0.029798001 0.6729099761 0.2972920233
[58,]  0.285156106 0.5825391456 0.1323047481
[59,]  0.031234714 0.7209584447 0.2478068412
[60,]  0.074746602 0.8305638837 0.0946895143
[61,]  0.218400944 0.6365414420 0.1450576136
[62,]  0.009185489 0.9620737604 0.0287407510
[63,]  0.055621313 0.8432649512 0.1011137362
[64,]  0.012118006 0.8995880340 0.0882939601
[65,]  0.091629644 0.8161955414 0.0921748145
[66,]  0.041721090 0.6898262780 0.2684526317
[67,]  0.014153805 0.9331611163 0.0526850788
[68,]  0.025884884 0.9257567915 0.0483583250
[69,]  0.027145071 0.8354092081 0.1374457207
[70,]  0.051544584 0.8776981932 0.0707572230
[71,]  0.027618640 0.7215665254 0.2508148344
[72,]  0.019490218 0.9342822228 0.0462275596
[73,]  0.023972824 0.7053570878 0.2706700884
[74,]  0.013936373 0.9027004542 0.0833631723
[75,]  0.022907805 0.8758831346 0.1012090602
[76,]  0.033942906 0.7547325843 0.2113245093
[77,]  0.033606150 0.5235844624 0.4428093876
[78,]  0.021184381 0.3062398204 0.6725757989
[79,]  0.004939666 0.9688476662 0.0262126679
[80,]  0.128301530 0.7670490155 0.1046494549
[81,]  0.077790647 0.8323328142 0.0898765391
[82,]  0.103762904 0.7955049785 0.1007321177
[83,]  0.030971559 0.9187363048 0.0502921366
[84,]  0.023972124 0.6562230039 0.3198048724
[85,]  0.026379608 0.8911646942 0.0824556978
[86,]  0.032144915 0.7971115685 0.1707435165
[87,]  0.033454318 0.5553331286 0.4112125531
[88,]  0.026864387 0.8576440693 0.1154915436
[89,]  0.024084133 0.9289403177 0.0469755498
[90,]  0.038167576 0.8994678822 0.0623645416
[91,]  0.019624492 0.9311594581 0.0492160495
[92,]  0.011621296 0.9156100918 0.0727686125
[93,]  0.022508771 0.9357165622 0.0417746670
[94,]  0.268911674 0.5983522463 0.1327360793
[95,]  0.012673718 0.9588977271 0.0284285546
[96,]  0.016778038 0.9455531803 0.0376687821
[97,]  0.009572210 0.9674237953 0.0230039948
[98,]  0.011408065 0.9438529899 0.0447389455
[99,]  0.355330531 0.5201593350 0.1245101338
[100,] 0.012683753   0.9605895752 0.0267266715
[101,] 0.019358235   0.1207293282 0.8599124372
[102,] 0.029297301   0.6155055693 0.3551971299
```

```
[103,] 0.006075879    0.0381501798   0.9557739416
[104,] 0.012522179    0.1418956202   0.8455822010
[105,] 0.004753887    0.0376327136   0.9576133993
[106,] 0.035459060    0.1526643550   0.8118765853
[107,] 0.072969289    0.7599903494   0.1670403613
[108,] 0.021897303    0.1151298020   0.8629728954
[109,] 0.013991215    0.1173156540   0.8686931314
[110,] 0.024416681    0.1145852161   0.8609981025
[111,] 0.016760627    0.2098109300   0.7734284433
[112,] 0.015751889    0.2230559309   0.7611921800
[113,] 0.001165779    0.0100110315   0.9888231890
[114,] 0.034364279    0.6599077515   0.3057279699
[115,] 0.038368189    0.4609331558   0.5006986550
[116,] 0.014060308    0.1360268918   0.8499127998
[117,] 0.007158380    0.0795776206   0.9132639993
[118,] 0.050587296    0.1858085498   0.7636041543
[119,] 0.049181070    0.1932444361   0.7575744942
[120,] 0.032187261    0.7106728992   0.2571398394
[121,] 0.003828058    0.0256996694   0.9704722729
[122,] 0.033640470    0.7069207083   0.2594388215
[123,] 0.041969706    0.1748179146   0.7832123794
[124,] 0.022801193    0.5956354579   0.3815633491
[125,] 0.002913570    0.0217413242   0.9753451062
[126,] 0.012907281    0.0751612211   0.9119314979
[127,] 0.020979914    0.7076636051   0.2713564807
[128,] 0.023091768    0.6467848930   0.3301233394
[129,] 0.008355500    0.0829718259   0.9086726743
[130,] 0.014444647    0.0947351575   0.8908201958
[131,] 0.019785664    0.1074380046   0.8727763312
[132,] 0.050902247    0.1889836824   0.7601140702
[133,] 0.008954187    0.0848145950   0.9062312176
[134,] 0.023389370    0.5400881802   0.4365224498
[135,] 0.031182190    0.3926627258   0.5761550847
[136,] 0.028660979    0.1317658493   0.8395731717
[137,] 0.017224871    0.1286753609   0.8540997684
[138,] 0.009780293    0.1100245651   0.8801951417
[139,] 0.021725820    0.7498049032   0.2284692770
[140,] 0.003488313    0.0289513662   0.9675603210
[141,] 0.005076956    0.0376725566   0.9572504872
[142,] 0.015399238    0.1294491591   0.8551516025
[143,] 0.029297301    0.6155055693   0.3551971299
[144,] 0.005250021    0.0337345741   0.9610154048
[145,] 0.009701055    0.0631749934   0.9271239514
[146,] 0.011259697    0.1063482072   0.8823920961
[147,] 0.025795992    0.5074231362   0.4667808722
[148,] 0.012109692    0.1563603928   0.8315299155
[149,] 0.021579041    0.1889929058   0.7894280532
[150,] 0.026920096    0.5816808796   0.3913990239
```

Closest hard clustering:

```
[1] 1 1 1 1 1 1 1 1 1 1 1 1   1 1 1 1 1 1 1 1 1 1 1 1 1 1 1 1 1 1 1 1 1 1 1 1
[37] 1 1 1 1 1 1 1 1 1 1 1 1   1 3 2 3 2 2 2 2 2 2 2 2 2 2 2 2 2 2 2 2 2 2 2 2
[73] 2 2 2 2 3 2 2 2 2 2 2 2   2 2 2 2 2 2 2 2 2 2 2 2 2 2 2 3 2 3 3 3 3 2 3
[109] 3 3 3 3 2 3 3 3 3 3 2 3 2 3 2 3 3 3 2 2 3 3 3 3 2 3 3 3 3 2 3 3 3 3 2 3
[145] 3 3 2 3 3 2
Available components:
[1] "centers"       "size"       "cluster"      "membership"   "iter"
[6] "withinerror"   "call"
```

调用 table() 函数显示聚类的结果，从以下结果得知有 13 个 versicolor 聚类到 virginica，有 3 个 virginica 聚类到 versicolor。

```
> table(iris$Species, result$cluster)

              1   2   3
  setosa     50   0   0
  versicolor  0  47   3
  virginica   0  13  37
```

7.4 聚类指标

由于使用非监督式的聚类算法时需先决定簇的数量，因此如何决定簇的数量便是一个很重要的工作。聚类指标可评估聚类的效果，协助用户决定簇的数量。

程序范例 7-4

先使用 NbClust 程序包和 iris 数据：

```
> library(NbClust)
> data(iris)
```

由于为无监督式学习，因此不使用 iris 数据中的目标属性（Species）：

```
> data <-iris[,-c(5)]
```

执行 NbClust()函数并设定自变量 distance="euclidean"、簇数目介于 min.nc=2 到 max.nc=6 之间、method ="kmeans"及 index="all"：

```
> NbClust(data, distance = "euclidean", min.nc=2, max.nc=6, method = "kmeans", index = "all")
*** : The Hubert index is a graphical method of determining the number of clusters.
In the plot of Hubert index, we seek a significant knee that corresponds to a
significant increase of the value of the measure i.e the significant peak in Hubert
index second differences plot.

*** : The D index is a graphical method of determining the number of clusters.
In the plot of D index, we seek a significant knee (the significant peak in Dindex
second differences plot) that corresponds to a significant increase of the value of
the measure.

All 150 observations were used.
```

```
******************************************************************
*   Among all indices:
*   8 proposed 2 as the best number of clusters
*   11 proposed 3 as the best number of clusters
*   1 proposed 4 as the best number of clusters
*   2 proposed 5 as the best number of clusters
*   2 proposed 6 as the best number of clusters

              ***** Conclusion *****
*According to the majority rule, the best number of clusters is  3

******************************************************************
$All.index
         KL        CH Hartigan      CCC    Scott  Marriot     TrCovW   TraceW
2    5.9068 513.9245 137.9491  35.9428 1044.605 467371.6  1045.9696 152.3480
3   12.4890 561.6278  15.2384  37.6701 1246.668 273408.6   248.9814  78.8514
4    9.7643 530.7658  20.5286  36.4682 1359.280 229428.4   173.8973  57.2285
5    0.8290 495.5415  -3.5725  35.2758 1465.533 176536.0   117.4449  46.4462
6   17.4419 441.7370  16.2762  33.6090 1527.121 168608.2    75.9611  41.7044
      Friedman     Rubin Cindex      DB Silhouette    Duda Pseudot2    Beale
2     732.8086   62.6152 0.2728  0.4744    0.6810  0.3153 186.7726   5.1885
3     801.6490  120.9780 0.3450  0.7256    0.5528  0.6779  27.5617   1.1284
4     874.3981  166.6878 0.3211  0.8436    0.4981  0.4309  63.3980   3.0858
5     989.3862  205.3837 0.2979  0.8571    0.4887  1.8245 -29.8255  -1.0623
6    1087.4777  228.7357 0.2850  0.9427    0.4841  2.3936 -19.7954  -1.3275
     Ratkowsky     Ball Ptbiserial    Frey McClain    Dunn  Hubert  SDindex  Dindex
2       0.5462  76.1740    0.8345  1.7571  0.2723  0.0765  0.0019   0.9995  0.8556
3       0.4967  26.2838    0.7146 -20.7325 0.5255  0.0988  0.0021   1.5740  0.6480
4       0.4413  14.3071    0.6361  5.5984  0.7120  0.1365  0.0021   2.3932  0.5574
5       0.3997   9.2892    0.6148 -4.4141  0.7671  0.0823  0.0022   2.7979  0.5097
6       0.3682   6.9507    0.5679  0.8351  0.9164  0.1089  0.0022   4.3487  0.4783
     SDbw
2    0.1618
3    0.2257
4    0.3186
5    0.0538
6    0.0469

$All.CriticalValues
  CritValue_Duda CritValue_PseudoT2 Fvalue_Beale
2         0.6357            49.2854       0.0004
3         0.5842            41.2743       0.3437
4         0.4837            51.2421       0.0185
5         0.5173            61.5845       1.0000
6         0.3773            56.1121       1.0000

$Best.nc
                KL       CH HartiganC     CC   Scott  Marriot   TrCovW
```

```
Number_clusters  6.0000      3.0000    3.0000   3.0000   3.0000      3.0    3.0000
Value_Index     17.4419    561.6278  122.7107  37.6701 202.0631 149982.9  796.9882
                 TraceW   Friedman     Rubin   Cindex        B Silhouette    Duda
Number_clusters  3.0000      5.000    5.0000   2.0000   2.0000      2.000  3.0000
Value_Index     51.8735    114.988  -15.3439   0.2728   0.4744      0.681  0.6779
                PseudoT2     Beale Ratkowsky     Ball PtBiserial     Frey McClain
Number_clusters  3.0000     3.0000    2.0000   3.0000   2.0000     2.0000  2.0000
Value_Index     27.5617     1.1284    0.5462  49.8902   0.8345     1.7571  0.2723
                  Dunn     Hubert   SDindex   Dindex     SDbw
Number_clusters  4.0000         0    2.0000        0   6.0000
Value_Index      0.1365         0    0.9995        0   0.0469

$Best.partition
  [1] 1 1 1 1 1 1 1 1 1 1 1 1 1 1 1 1 1 1 1 1 1 1 1 1 1 1 1 1 1 1 1 1 1 1 1 1
 [37] 1 1 1 1 1 1 1 1 1 1 1 1 1 1 2 2 3 2 2 2 2 2 2 2 2 2 2 2 2 2 2 2 2 2 2 2
 [73] 2 2 2 2 2 3 2 2 2 2 2 2 2 2 2 2 2 2 2 2 2 2 2 2 2 2 3 3 3 3 3 2 3
[109] 3 3 3 3 3 2 2 3 3 3 3 2 3 2 3 3 3 2 2 3 3 3 3 3 2 3 3 3 3 2 3 3 3 2 3
[145] 3 3 2 3 3 2
```

从以上数据显示，有11种聚类指标表示使用3簇是最好的簇数目。

第 8 章 进化式学习

进化式学习是指模拟自然界进化过程所建立的学习模型，通常将进化式学习所建立的模型称为进化式算法。在本章中，我们将介绍常用的进化式算法，包含基因算法和人工蜂群算法。

本章重点内容：

- 基因算法
- 人工蜂群算法

8.1 基因算法

Holland 等人于 1975 年提出了基因算法（或称为遗传算法）的基本理论，他的基本精神在于模仿自然界中的物竞天择、优胜劣败的自然进化规则，选择物种对环境适应力较强的亲代（Parents Generation），并随机相互交换彼此的基因信息，以期产生更优秀的子代（Offspring Generation），经过筛选（Selection）后留下适应力最优的物种，再继续组合交叉、繁衍及筛选，如此重复不断地进化出对外部环境适应力最强的物种。

复制（Reproduction）、组合交叉（Crossover）、变异（Mutation）是基因算法基本的三种运算机制。基因算法是将问题通过编码的方式转换到染色体（个体）结构上，然后使用适应度函数（Fitness Function）的定义评估（Evaluation）优劣程度，经由复制、组合交叉、变异等运算机制产生新的染色体（Chromosome），并取代（Replace）在族群中表现不好的染色体。进化学习的过程是不断反复地评估、筛选，找到适应度函数最优的染色体以保留到下一代（Generation）、继续进化，直到满足终止条件为止。基因算法基本概念如下：

（1）将问题以编码的方式对应到一个解（Solution），在基因算法里被称为染色体或个体（Individual）。

（2）使用并行搜索，同时产生多组染色体来进行随机搜索，在基因算法里被称为族群（Population）。

（3）根据染色体的适应度函数评估解的优劣。

（4）通过进化复制的过程，使用随机的方式将上一代的部分基因移转到下一代身上，创造出不同于上一代的新染色体，称之为基因运算机制。

基因算法的流程如图 8-1 所示。

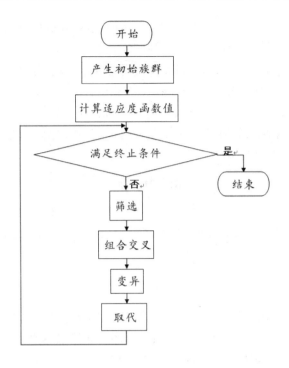

图 8-1 基因算法的流程图

基因算法求解的第一个步骤为产生初始族群。在使用二进制基因算法时，可先将所要求解的优化问题参数加以编码，编码为一定长度的二进制字符串，例如：

```
0110101101
1100011000
```

若解为实数，则可使用公式 (8-1) 将二进制转换为实数。

$$x = B2D * (UB - LB) / (2^L - 1) + LB \tag{8-1}$$

其中：

- B2D 二进制转换成十进制数。
- UB 上界。
- LB 下界。
- L 二进制字符串长度。
- X 输出实数值。

基因算法有三种主要的运算机制，分别为复制、组合交叉及变异。首先，在进行复制的运算之前必须先进行选择（或筛选）。选择的方法很多，例如轮盘赌式选择（Roulette Wheel Selection）与锦标赛式选择（Tournament Selection）。以轮盘赌式选择为例，这种选择是根据每个染色体的适应度函数值来决定该染色体在轮盘上的面积，适应度函数值越大面积就越大，其被选择到的概率就越大，被选到的染色体会被放到组合交叉池中等待组合交叉。锦标赛式选择就是在每一代的进化过程中随机选择两个或多个染色体，具有最大适应度函数值的染色体即被选中送到组合交叉池中。进行组合交叉的过程中，必须先决定组合交叉概率的大小（Crossover Probability），每一对染色体都将按照组合交叉概率来决定是否进行组合交叉，希望通过组合交叉过程能持续累积优秀的染色体，使得组合交叉产生的子代适应度函数值越来越高。常用的组合交叉方法有单点组合交叉（One-Point Crossover）、双点组合交叉（Two-Point Crossover）、多点组合交叉（Multi-point Crossover）、掩码组合交叉（Mask Crossover）及概率均等式组合交叉（Uniform Crossover，或称为均匀组合交叉）等。单点组合交叉将族群内随机选取的两段基因彼此交换，而组成两个新的基因字符串。假设算法选择 A 的第 5 位（比特）作为组合交叉点（从右边算起）：

$$A=0110011111$$
$$B=1100100000$$

组合交叉点选择完毕之后，接着将两个染色体中位于组合交叉点后的所有位互换，结果如下：

$$A=0110000000$$
$$B=1100111111$$

组合交叉完成后，A 进化成为 0110000000，而 B 进化成为 1100111111。通过复制与组合交叉会产生许多不同的新染色体，单以染色体上的基因来说，未产生新的信息。因此，希望能通过某些方式使基因产生新的信息，这种方法就是变异，如同先前的组合交叉过程，变异过程也需设定一个变异概率（Mutation Probability）。简单的变异运算方式可选取任意一个染色体中的某一个基因，再将该点所代表的位取反，例如：

$$A=010000000$$

假设算法选择 A 的第 8 位（比特）作为变异点，则：

$$A=010000000 \rightarrow A=000000000$$

变异完成后，A 从 010000000 变异为 000000000。取代部分以精英政策为主，其目的是将旧族群中适应度函数值最优者与前几名的染色体存留下来，其余的以子代新染色体取代，成为新的族群。例如，当精英政策设定为 0.6 时，代表有 60% 亲代的染色体保留下来（来自前代族群中适应度函数值的前 60% 的染色体），而其余的 40% 则用表现较佳的子代染色体来取代，成为新的族群。每一代都会重复上述操作，不断地更新族群中的染色体，通过精英政策的方式来留住表现较佳的染色体，直到满足终止条件为止。

程序范例 8-1

首先使用 GA 程序包：

```
> library("GA")
```

本范例中欲求 Max.$f(x) = 25 - x * x$, $-5 \leqslant x \leqslant 5$：

```
> f <- function(x)  25-x*x
> min <- -5
> max <- +5
```

使用 curve() 函数绘出 Max.$f(x) = 25 - x * x$, $-5 \leq x \leq 5$ 的图，如图 8-2 所示。

```
> curve(f, min, max)
```

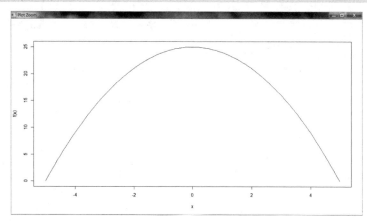

图 8-2　$f(x) = 25 - x*x$ 的图

设置适应函数为 $f(x)$：

```
> fitness <- function(x) f(x)
```

执行 ga() 函数并设置自变量 popSize= 50、pcrossover= 0.8、pmutation= 0.1、elitism=10、monitor = gaMonitor、maxiter=100，最好的解为 $f(x) = 25$：

```
> GA <- ga(type="real-valued",
+ fitness=fitness,
+ min=min,
+ max=max,
+ popSize = 50,
+ pcrossover = 0.8,
+ pmutation = 0.1,
+ elitism = 10,
+ monitor = gaMonitor,
+ maxiter = 100)

Iter = 1  | Mean = 16.75026 | Best = 24.99012
Iter = 2  | Mean = 22.87244 | Best = 24.99869
Iter = 3  | Mean = 23.82392 | Best = 24.99869
Iter = 4  | Mean = 23.77883 | Best = 24.99869
Iter = 5  | Mean = 23.25862 | Best = 24.99999
Iter = 6  | Mean = 23.89032 | Best = 24.99999
Iter = 7  | Mean = 24.16965 | Best = 24.99999
Iter = 8  | Mean = 24.31778 | Best = 25
Iter = 9  | Mean = 24.04458 | Best = 25
Iter = 10 | Mean = 24.19527 | Best = 25
Iter = 11 | Mean = 24.66865 | Best = 25
Iter = 12 | Mean = 24.76368 | Best = 25
```

```
Iter = 13  | Mean = 23.61497 | Best = 25
Iter = 14  | Mean = 24.51597 | Best = 25
Iter = 15  | Mean = 24.84668 | Best = 25
Iter = 16  | Mean = 24.70468 | Best = 25
Iter = 17  | Mean = 23.72271 | Best = 25
Iter = 18  | Mean = 23.83589 | Best = 25
Iter = 19  | Mean = 23.83062 | Best = 25
Iter = 20  | Mean = 24.88215 | Best = 25
Iter = 21  | Mean = 23.84913 | Best = 25
Iter = 22  | Mean = 24.12524 | Best = 25
Iter = 23  | Mean = 24.00631 | Best = 25
Iter = 24  | Mean = 24.24212 | Best = 25
Iter = 25  | Mean = 24.33541 | Best = 25
Iter = 26  | Mean = 24.6339  | Best = 25
Iter = 27  | Mean = 24.40812 | Best = 25
Iter = 28  | Mean = 24.80727 | Best = 25
Iter = 29  | Mean = 23.39615 | Best = 25
Iter = 30  | Mean = 24.64121 | Best = 25
Iter = 31  | Mean = 23.97989 | Best = 25
Iter = 32  | Mean = 23.76933 | Best = 25
Iter = 33  | Mean = 23.93873 | Best = 25
Iter = 34  | Mean = 23.98081 | Best = 25
Iter = 35  | Mean = 24.16118 | Best = 25
Iter = 36  | Mean = 24.31206 | Best = 25
Iter = 37  | Mean = 23.61612 | Best = 25
Iter = 38  | Mean = 24.56559 | Best = 25
Iter = 39  | Mean = 24.53877 | Best = 25
Iter = 40  | Mean = 23.64468 | Best = 25
Iter = 41  | Mean = 24.3073  | Best = 25
Iter = 42  | Mean = 24.52696 | Best = 25
Iter = 43  | Mean = 24.73332 | Best = 25
Iter = 44  | Mean = 23.64634 | Best = 25
Iter = 45  | Mean = 24.46134 | Best = 25
Iter = 46  | Mean = 24.07619 | Best = 25
Iter = 47  | Mean = 23.87074 | Best = 25
Iter = 48  | Mean = 24.18712 | Best = 25
Iter = 49  | Mean = 24.33388 | Best = 25
Iter = 50  | Mean = 24.12963 | Best = 25
Iter = 51  | Mean = 24.66332 | Best = 25
Iter = 52  | Mean = 24.45472 | Best = 25
Iter = 53  | Mean = 23.33875 | Best = 25
Iter = 54  | Mean = 24.32067 | Best = 25
Iter = 55  | Mean = 24.01558 | Best = 25
Iter = 56  | Mean = 23.85615 | Best = 25
Iter = 57  | Mean = 24.2742  | Best = 25
Iter = 58  | Mean = 24.4183  | Best = 25
```

```
Iter = 59  | Mean = 23.67884 | Best = 25
Iter = 60  | Mean = 23.69734 | Best = 25
Iter = 61  | Mean = 23.4972  | Best = 25
Iter = 62  | Mean = 24.9315  | Best = 25
Iter = 63  | Mean = 24.43032 | Best = 25
Iter = 64  | Mean = 22.74806 | Best = 25
Iter = 65  | Mean = 24.14798 | Best = 25
Iter = 66  | Mean = 23.36102 | Best = 25
Iter = 67  | Mean = 24.2989  | Best = 25
Iter = 68  | Mean = 24.67911 | Best = 25
Iter = 69  | Mean = 23.88688 | Best = 25
Iter = 70  | Mean = 24.46076 | Best = 25
Iter = 71  | Mean = 24.44833 | Best = 25
Iter = 72  | Mean = 24.1129  | Best = 25
Iter = 73  | Mean = 23.45282 | Best = 25
Iter = 74  | Mean = 24.12345 | Best = 25
Iter = 75  | Mean = 23.4959  | Best = 25
Iter = 76  | Mean = 23.90719 | Best = 25
Iter = 77  | Mean = 24.46473 | Best = 25
Iter = 78  | Mean = 24.51689 | Best = 25
Iter = 79  | Mean = 23.56051 | Best = 25
Iter = 80  | Mean = 23.99792 | Best = 25
Iter = 81  | Mean = 24.5327  | Best = 25
Iter = 82  | Mean = 22.67488 | Best = 25
Iter = 83  | Mean = 24.0043  | Best = 25
Iter = 84  | Mean = 23.64904 | Best = 25
Iter = 85  | Mean = 24.47736 | Best = 25
Iter = 86  | Mean = 23.67569 | Best = 25
Iter = 87  | Mean = 23.31366 | Best = 25
Iter = 88  | Mean = 23.93136 | Best = 25
Iter = 89  | Mean = 23.81057 | Best = 25
Iter = 90  | Mean = 23.2838  | Best = 25
Iter = 91  | Mean = 24.3743  | Best = 25
Iter = 92  | Mean = 24.27372 | Best = 25
Iter = 93  | Mean = 22.63764 | Best = 25
Iter = 94  | Mean = 23.6324  | Best = 25
Iter = 95  | Mean = 24.77453 | Best = 25
Iter = 96  | Mean = 24.48018 | Best = 25
Iter = 97  | Mean = 23.78382 | Best = 25
Iter = 98  | Mean = 23.28074 | Best = 25
Iter = 99  | Mean = 23.29128 | Best = 25
Iter = 100 | Mean = 24.03805 | Best = 25
```

使用 plot() 函数画出执行的结果，如图 8-3 所示。

```
> plot(GA)
```

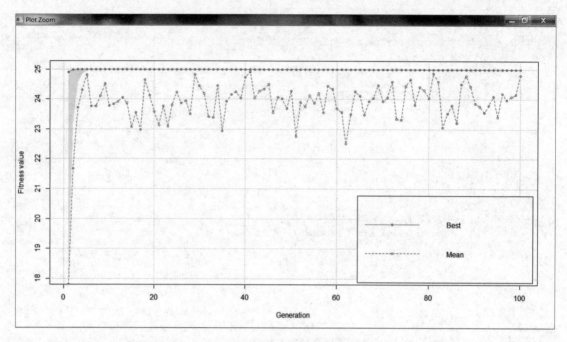

图 8-3　基因算法执行的结果

8.2　人工蜂群算法

人工蜂群算法建立在蜜蜂群体智慧的基础上，其核心是由觅食蜂（The Employed Bee）、跟随蜂（The Onlooker Bee）、侦察蜂（The Scout Bee）以及食物源（Sources）所组成的。蜜蜂对食物源的搜索由三步组成：觅食蜂发现食物源并记录下花蜜的数量；跟随蜂根据觅食蜂所提供的花蜜信息来选定到哪个食物源采蜜；侦察蜂随机搜索蜂巢附近的食物源以寻找新的食物源。食物源相当于优化问题中解的位置，在人工蜂群算法中食物源（解）的价值由适应度（Fitness）值来表示。某个解连续经过有限次循环之后而没有得到改善，表示其陷入了局部最优解，则这个食物源就要被放弃，放弃所采蜜（花粉）食物源的觅食蜂将成为侦察蜂，并随机搜索一个新的食物源。

Karaboga 提出了人工蜂群（Artificial Bee Colony，ABC）算法并成功地将其应用在函数数值优化问题上。在人工蜂群算法中有一半的蜜蜂是觅食蜂，另一半是跟随蜂。每一处食物源仅有一只觅食蜂，也就是说食物源和觅食蜂的数量相等。觅食蜂存储某一个食物源的相关数据（位置及花蜜数量等），并将这些信息以一定的概率与其他蜜蜂分享。觅食蜂在蜂巢内将它们的信息通过舞蹈传递给跟随蜂，跟随蜂随机选择一个食物源并变为觅食蜂。在人工蜂群算法中，跟随蜂选择食物源的概率（Pi）根据公式 (8-2) 计算：

$$P_i = \frac{Fit_i}{\sum_{i=1}^{SN} Fit_i} \tag{8-2}$$

其中：

- SN 表示食物源的数量。

- Fit_i 表示第 i 个食物源的适应度值。

跟随蜂根据公式 (8-3) 更新位置：

$$v_{ij}(t+1) = x_{ij}(t) + \varnothing \ (x_{ij}(t) - x_{kj}(t)) \tag{8-3}$$

其中：

- V_{ij} 表示第 i 个跟随蜂在第 j 维空间的食物源的位置。
- t ：表示迭代数。
- X_i 表示一个代表第 i 个食物源的向量且 $X_i=(x_{i1}, x_{i2}, ..., x_{id})$。
- x_{kj} 表示第 k 个食物源的位置，k 是随机选取的，$k = \text{int}(rand*SN) + 1$。
- ϕ 表示一个随机值，介于 $[-1,1]$。

侦察蜂随机搜索一个新的食物源时，根据公式 (8-4) 更新位置：

$$x_{ij} = x_{\min}^{j} + \gamma * \left(x_{\max}^{j} - x_{\min}^{j}\right) \tag{8-4}$$

其中：

- γ 表示一个介于 $[0, 1]$ 的随机值。
- $x_{\min}$ 表示位于 j 维空间食物源位置的下界（Lower Bound）。
- $x_{\max}$ 表示位于 j 维空间食物源位置的上界（Upper Bound）。

程序范例 8-2

首先使用 ABCoptim 程序包：

```
> library(ABCoptim)
```

本范例中欲求 $\text{Min}.f(x) = -\cos x_1 * \cos x_2 * \exp(-((x_1-\pi)^2+(x_2-\pi)^2))$：

```
> fun <- function(x) {
+   -cos(x[1])*cos(x[2])*exp(-((x[1] - pi)^2 + (x[2] - pi)^2))
+ }
```

调用 abc_optim() 函数并设置自变量，其中调用 rep(0, 2) 函数产生的初始值为（0，0）、解的范围为 -20 ≤ x1, x2 ≤ 20，迭代次数 criter=1000：

```
> abc_optim(rep(0,2), fun, lb=-20, ub=20, criter=1000)
$par
[1] 3.141593 3.141593

$value
[1] -1

$counts
function
    1000
```

由以上结果得知，$x1, x2 = 3.141593$ 时，$f(\underline{x}) = -1$。

程序范例 8-3

首先使用 ABCoptim 程序包：

```
> library(ABCoptim)
```

本范例中欲求 $Min.f(x) = 10*sin(0.3*x)*sin(1.3*x^2) + 0.00001*x^4 + 0.2*x + 80$：

```
> fw <- function (x)
+    10*sin(0.3*x)*sin(1.3*x^2) + 0.00001*x^4 + 0.2*x+80
```

调用 abc_optim() 函数并设置自变量，设置初始值为 50、解的范围为 $-100 \leq x_1, x_2 \leq 100$，迭代次数 criter=1000：

```
> abc_optim(50, fw, lb=-100, ub=100, criter=1000)
$par
[1] -15.81515
$value
[1] 67.46773

$counts
function
    1000
```

由以上结果得知，$x = -15.81515$ 时，$f(x) = 67.46773$。

第 9 章 混合式学习

混合式学习是综合两种以上学习法的优点以提升单一学习法的性能或效率。使用 C 5.0 决策树时，最少案例数量（minCases）、修剪树置信水平两项参数在面对不同问题时会有不同的最优参数组合，因此可使用人工蜂群算法来调整决策树参数，以取得较好的分类正确率。

本章重点内容：
- 使用 C50 和 ABCoptim 程序包范例
- 使用基因算法来调整人工神经网络参数范例

9.1 使用 C50 和 ABCoptim 程序包范例

程序范例 9-1

首先使用 C50 和 ABCoptim 程序包：

```
> library(C50)
> library(ABCoptim)
```

创建自定义函数 test.error()，其输入自变量 xx[1] 为 CF 值（1）、xx[2] 为 minCases 值（minCases >=2），输出值为测试数据的错误率。本范例中以 best_CF、best_minCases 及 min_error 来记录最优的 CF、minCases 及测试数据的最小错误率，注意必须使用 <<- 才能改变自定义函数外部对象的值。

```
> test.error <- function(xx){
+   c <- C5.0Control(subset = FALSE,
+                    bands = 0,
+                    winnow = FALSE,
+                    noGlobalPruning = FALSE,
+                    CF = xx[1]/100,
+                    minCases = floor(xx[2]),
+                    fuzzyThreshold = FALSE,
+                    sample = 0,   # for holdout
+                    seed = sample.int(4096, size = 1) - 1L,
+                    earlyStopping = TRUE
+   )
+
+   treeModel <- C5.0(x = iris.train[, -5], y = iris.train$Species,control =c)
+   summary(treeModel)
+
```

```
+     test.output <- predict(treeModel, iris.test[, -5], type = "class")
+
+     n=length(test.output)
+     number=0
+     for( i in 1:n)
+     {
+       if (test.output[i] != iris.test[i,5])
+       {
+        number=number+1 # Error
+       }
+     }
+
+     error_value <- number/n*100
+
+     if (start_index == TRUE)
+     {
+       best_CF <<- c$CF # Keep best global parameters
+       best_minCases <<- c$minCases
+       min_error <<- error_value
+     }
+
+     if (error_value < min_error )
+     {
+       best_CF <<- c$CF # Keep best global parameters
+       best_minCases <<- c$minCases
+       min_error <<- error_value
+     }
+
+     error=error_value
+     return(error)
+ }
```

在主程序中使用 iris 数据：

```
>##############################################################
> #  Main program
>##############################################################
> data(iris)
```

使用 50% 为测试数据：

```
> np = ceiling(0.5*nrow(iris))
> np
[1] 75
```

区分测试数据 iris.test 和训练数据 iris.train：

```
> iris.test = iris[1:np,]
> iris.train = iris[np+1:nrow(iris),]
```

先以初始值 CF=85%、minCases=25 执行 C5.0 决策树并返回分类错误率为 68%：

```
> start_index <<- TRUE           # Global value
```

```
> xx=c(85,2)                    # Initial values
> test.error(xx)
[1] 68
```

接着调用 abc_optim() 函数，使用相同初始值、解的范围，最大迭代次数 maxCycle = 10。执行后分类错误率由 68% 降至 66.66667%。

```
> start_index <<- FALSE         # Global value
> abc_optim(xx,test.error,lb=2, ub=100, maxCycle=10)
$par
[1] 32.26316 32.26316

$value
[1] 66.66667

$counts
function
      10
> accuracy=100-min_error
> accuracy
[1] 33.33333
```

用户也可使用基因算法来调整人工神经网络参数，以得到更好的效果。ANN 程序包中提供了 ANNGA() 函数和演示（Demo）范例。

9.2 使用基因算法来调整人工神经网络参数的范例

程序范例 9-2

```
> data("dataANN")
>
> ANNGA(x =input,
+ y =output,
+ design =c(1, 3, 1),
+ population =100,
+ mutation = 0.3,
+ crossover = 0.7,
+ maxGen =100,
+ error =0.001)

***cycle***
Generation:  1  Best population fitness : 0.05981754 Mean of population:0.25887034
Best chromosome->-20.36/13.89/1.50/-23.98/-9.71/9.35/15.48/-14.58/0.68/-15.66/

***cycle***
Generation:  2  Best population fitness : 0.03989556 Mean of  population:0.23688910
  Best chromosome->32.96/18.16/-2.28/-30.37/3.01/-16.76/0.88/21.82/13.42/-1.07/
```

```
***cycle***
Generation:  3  Best population fitness : 0.03989556 Mean of population:0.22232952
 Best chromosome->32.96/18.16/-2.28/-30.37/3.01/-16.76/0.88/21.82/13.42/-1.07/

***cycle***
Generation:  4  Best population fitness : 0.03880315 Mean of population:0.20310358
 Best chromosome->-0.51/15.81/18.32/15.46/-16.12/-6.78/3.00/8.42/35.82/-11.58/

***cycle***
Generation:  5  Best population fitness : 0.03880315 Mean of population:0.18924708
 Best chromosome->-0.51/15.81/18.32/15.46/-16.12/-6.78/3.00/8.42/35.82/-11.58/

***cycle***
Generation:  6  Best population fitness : 0.03789907 Mean of population:0.17406013
 Best chromosome->-23.55/-25.39/8.95/14.06/19.25/11.80/-9.81/-2.75/17.90/-15.04/

***cycle***
Generation:  7  Best population fitness : 0.03789907 Mean of population:0.16399078
 Best chromosome->-23.55/-25.39/8.95/14.06/19.25/11.80/-9.81/-2.75/17.90/-15.04/

***cycle***
Generation:  8  Best population fitness : 0.03789907 Mean of population:0.15730231
 Best chromosome->-23.55/-25.39/8.95/14.06/19.25/11.80/-9.81/-2.75/17.90/-15.04/

***cycle***
Generation:  9  Best population fitness : 0.03789907 Mean of population:0.14605940
 Best chromosome->-23.55/-25.39/8.95/14.06/19.25/11.80/-9.81/-2.75/17.90/-15.04/

***cycle***
Generation:  10  Best population fitness : 0.03789907 Mean of population:0.13908024
 Best chromosome->-23.55/-25.39/8.95/14.06/19.25/11.80/-9.81/-2.75/17.90/-15.04/

***cycle***
Generation:  11  Best population fitness : 0.03789907 Mean of population:0.13207254
 Best chromosome->-23.55/-25.39/8.95/14.06/19.25/11.80/-9.81/-2.75/17.90/-15.04/

***cycle***
Generation:  12  Best population fitness : 0.03789907 Mean of population:0.12698297
 Best chromosome->-23.55/-25.39/8.95/14.06/19.25/11.80/-9.81/-2.75/17.90/-15.04/

***cycle***
Generation:  13  Best population fitness : 0.03789907 Mean of population:0.12094905
 Best chromosome->-23.55/-25.39/8.95/14.06/19.25/11.80/-9.81/-2.75/17.90/-15.04/

***cycle***
Generation:  14  Best population fitness : 0.03789907 Mean of population:0.11631263
 Best chromosome->-23.55/-25.39/8.95/14.06/19.25/11.80/-9.81/-2.75/17.90/-15.04/

***cycle***
Generation:  15  Best population fitness : 0.03697816 Mean of population:0.11132284
```

Best chromosome->14.83/-15.09/34.34/8.66/-7.20/-20.87/-4.18/0.87/-5.00/-0.78/

cycle
Generation: 16 Best population fitness : 0.03647556 Mean of population:0.10676000
 Best chromosome->14.35/-15.80/-1.29/-13.05/10.05/-20.28/-4.23/21.93/-0.11/-0.03/

cycle
Generation: 17 Best population fitness : 0.03647556 Mean of population:0.10152121
 Best chromosome->14.35/-15.80/-1.29/-13.05/10.05/-20.28/-4.23/21.93/-0.11/-0.03/

cycle
Generation: 18 Best population fitness : 0.03647556 Mean of population:0.09759118
 Best chromosome->14.35/-15.80/-1.29/-13.05/10.05/-20.28/-4.23/21.93/-0.11/-0.03/

cycle
Generation: 19 Best population fitness : 0.03647556 Mean of population:0.09328037
 Best chromosome->14.35/-15.80/-1.29/-13.05/10.05/-20.28/-4.23/21.93/-0.11/-0.03/

cycle
Generation: 20 Best population fitness : 0.03647556 Mean of population:0.09093867
 Best chromosome->14.35/-15.80/-1.29/-13.05/10.05/-20.28/-4.23/21.93/-0.11/-0.03/

cycle
Generation: 21 Best population fitness : 0.03647556 Mean of population:0.08929980
 Best chromosome->14.35/-15.80/-1.29/-13.05/10.05/-20.28/-4.23/21.93/-0.11/-0.03/

cycle
Generation: 22 Best population fitness : 0.03647556 Mean of population:0.08791822
 Best chromosome->14.35/-15.80/-1.29/-13.05/10.05/-20.28/-4.23/21.93/-0.11/-0.03/

cycle
Generation: 23 Best population fitness : 0.03647556 Mean of population:0.08586487
 Best chromosome->14.35/-15.80/-1.29/-13.05/10.05/-20.28/-4.23/21.93/-0.11/-0.03/

cycle
Generation: 24 Best population fitness : 0.03647556 Mean of population:0.08391625
 Best chromosome->14.35/-15.80/-1.29/-13.05/10.05/-20.28/-4.23/21.93/-0.11/-0.03/

cycle
Generation: 25 Best population fitness : 0.03538457 Mean of population:0.08210314
 Best chromosome->14.83/-23.51/3.32/8.66/12.33/-13.19/-4.18/0.87/-5.00/-0.78/

cycle
Generation: 26 Best population fitness : 0.03538457 Mean of population:0.07831619
 Best chromosome->14.83/-23.51/3.32/8.66/12.33/-13.19/-4.18/0.87/-5.00/-0.78/

cycle
Generation: 27 Best population fitness : 0.03538457 Mean of population:0.07424434
 Best chromosome->14.83/-23.51/3.32/8.66/12.33/-13.19/-4.18/0.87/-5.00/-0.78/

```
***cycle***
Generation:  28  Best population fitness : 0.03538457 Mean of population:0.07256437
 Best chromosome->14.83/-23.51/3.32/8.66/12.33/-13.19/-4.18/0.87/-5.00/-0.78/

***cycle***
Generation:  29  Best population fitness : 0.03538457 Mean of population:0.07032876
 Best chromosome->14.83/-23.51/3.32/8.66/12.33/-13.19/-4.18/0.87/-5.00/-0.78/

***cycle***
Generation:  30  Best population fitness : 0.03538457 Mean of population:0.06901380
 Best chromosome->14.83/-23.51/3.32/8.66/12.33/-13.19/-4.18/0.87/-5.00/-0.78/

***cycle***
Generation:  31  Best population fitness : 0.03538457 Mean of population:0.06769783
 Best chromosome->14.83/-23.51/3.32/8.66/12.33/-13.19/-4.18/0.87/-5.00/-0.78/

***cycle***
Generation:  32  Best population fitness : 0.03538457 Mean of population:0.06589963
 Best chromosome->14.83/-23.51/3.32/8.66/12.33/-13.19/-4.18/0.87/-5.00/-0.78/

***cycle***
Generation:  33  Best population fitness : 0.03538457 Mean of population:0.06435588
 Best chromosome->14.83/-23.51/3.32/8.66/12.33/-13.19/-4.18/0.87/-5.00/-0.78/

***cycle***
Generation:  34  Best population fitness : 0.03538457 Mean of population:0.06406965
 Best chromosome->14.83/-23.51/3.32/8.66/12.33/-13.19/-4.18/0.87/-5.00/-0.78/

***cycle***
Generation:  35  Best population fitness : 0.03538457 Mean of population:0.06037107
 Best chromosome->14.83/-23.51/3.32/8.66/12.33/-13.19/-4.18/0.87/-5.00/-0.78/

***cycle***
Generation:  36  Best population fitness : 0.03538457 Mean of population:0.05790038
 Best chromosome->14.83/-23.51/3.32/8.66/12.33/-13.19/-4.18/0.87/-5.00/-0.78/

***cycle***
Generation:  37  Best population fitness : 0.03538457 Mean of population:0.05684195
 Best chromosome->14.83/-23.51/3.32/8.66/12.33/-13.19/-4.18/0.87/-5.00/-0.78/

***cycle***
Generation:  38  Best population fitness : 0.03538457 Mean of population:0.05615396
 Best chromosome->14.83/-23.51/3.32/8.66/12.33/-13.19/-4.18/0.87/-5.00/-0.78/

***cycle***
Generation:  39  Best population fitness : 0.03538457 Mean of population:0.05344059
 Best chromosome->14.83/-23.51/3.32/8.66/12.33/-13.19/-4.18/0.87/-5.00/-0.78/

***cycle***
Generation:  40  Best population fitness : 0.03538457 Mean of population:0.05124761
```

Best chromosome->14.83/-23.51/3.32/8.66/12.33/-13.19/-4.18/0.87/-5.00/-0.78/

cycle
Generation: 41 Best population fitness : 0.03538457 Mean of population:0.05065232
Best chromosome->14.83/-23.51/3.32/8.66/12.33/-13.19/-4.18/0.87/-5.00/-0.78/

cycle
Generation: 42 Best population fitness : 0.03431974 Mean of population:0.04800988
Best chromosome->17.69/-13.49/34.59/-61.31/-2.44/-23.15/-0.19/2.56/16.33/0.09/

cycle
Generation: 43 Best population fitness : 0.03431974 Mean of population:0.04658839
Best chromosome->17.69/-13.49/34.59/-61.31/-2.44/-23.15/-0.19/2.56/16.33/0.09/

cycle
Generation: 44 Best population fitness : 0.03431974 Mean of population:0.04567413
Best chromosome->17.69/-13.49/34.59/-61.31/-2.44/-23.15/-0.19/2.56/16.33/0.09/

cycle
Generation: 45 Best population fitness : 0.03350304 Mean of population:0.04370310
Best chromosome->10.08/-13.20/-51.64/-43.34/14.68/-18.22/-52.58/8.05/28.31/0.14/

cycle
Generation: 46 Best population fitness : 0.03350304 Mean of population:0.04300280
Best chromosome->10.08/-13.20/-51.64/-43.34/14.68/-18.22/-52.58/8.05/28.31/0.14/

cycle
Generation: 47 Best population fitness : 0.03350304 Mean of population:0.04107488
Best chromosome->10.08/-13.20/-51.64/-43.34/14.68/-18.22/-52.58/8.05/28.31/0.14/

cycle
Generation: 48 Best population fitness : 0.03350304 Mean of population:0.04033673
Best chromosome->10.08/-13.20/-51.64/-43.34/14.68/-18.22/-52.58/8.05/28.31/0.14/

cycle
Generation: 49 Best population fitness : 0.03350304 Mean of population:0.03908786
Best chromosome->10.08/-13.20/-51.64/-43.34/14.68/-18.22/-52.58/8.05/28.31/0.14/

cycle
Generation: 50 Best population fitness : 0.03350304 Mean of population:0.03899032
Best chromosome->10.08/-13.20/-51.64/-43.34/14.68/-18.22/-52.58/8.05/28.31/0.14/

cycle
Generation: 51 Best population fitness : 0.03350304 Mean of population:0.03897411
Best chromosome->10.08/-13.20/-51.64/-43.34/14.68/-18.22/-52.58/8.05/28.31/0.14/

cycle
Generation: 52 Best population fitness : 0.03350304 Mean of population:0.03888024
Best chromosome->10.08/-13.20/-51.64/-43.34/14.68/-18.22/-52.58/8.05/28.31/0.14/

```
***cycle***
Generation: 53  Best population fitness : 0.03350304 Mean of population:0.03876724
 Best chromosome->10.08/-13.20/-51.64/-43.34/14.68/-18.22/-52.58/8.05/28.31/0.14/

***cycle***
Generation: 54  Best population fitness : 0.03350304 Mean of population:0.03831313
 Best chromosome->10.08/-13.20/-51.64/-43.34/14.68/-18.22/-52.58/8.05/28.31/0.14/

***cycle***
Generation: 55  Best population fitness : 0.03341240 Mean of population:0.03777473
 Best chromosome->13.88/-16.12/14.95/-32.53/-6.31/-3.61/-16.03/30.22/19.17/0.03/

***cycle***
Generation: 56  Best population fitness : 0.03341240 Mean of population:0.03764631
 Best chromosome->13.88/-16.12/14.95/-32.53/-6.31/-3.61/-16.03/30.22/19.17/0.03/

***cycle***
Generation: 57  Best population fitness : 0.03227431 Mean of population:0.03724999
 Best chromosome->10.90/-42.97/-30.51/-65.17/3.04/-7.17/-48.00/8.93/-25.48/0.10/

***cycle***
Generation: 58  Best population fitness : 0.03227431 Mean of population:0.03720404
 Best chromosome->10.90/-42.97/-30.51/-65.17/3.04/-7.17/-48.00/8.93/-25.48/0.10/

***cycle***
Generation: 59  Best population fitness : 0.03227431 Mean of population:0.03711149
 Best chromosome->10.90/-42.97/-30.51/-65.17/3.04/-7.17/-48.00/8.93/-25.48/0.10/

***cycle***
Generation: 60  Best population fitness : 0.03227431 Mean of population:0.03647350
 Best chromosome->10.90/-42.97/-30.51/-65.17/3.04/-7.17/-48.00/8.93/-25.48/0.10/

***cycle***
Generation: 61  Best population fitness : 0.03177191 Mean of population:0.03634610
 Best chromosome->16.25/-12.34/16.80/-46.88/-30.21/-22.08/-1.62/19.55/10.75/0.09/

***cycle***
Generation: 62  Best population fitness : 0.02853393 Mean of population:0.03619104
 Best chromosome->-20.36/-21.39/23.15/-39.82/9.42/-4.25/-4.70/0.95/-0.41/0.11/

***cycle***
Generation: 63  Best population fitness : 0.02428789 Mean of population:0.03595940
 Best chromosome->10.70/-19.86/36.83/-20.32/18.88/-39.12/-40.71/-0.97/-23.64/0.08/

***cycle***
Generation: 64  Best population fitness : 0.02428789 Mean of population:0.03590585
 Best chromosome->10.70/-19.86/36.83/-20.32/18.88/-39.12/-40.71/-0.97/-23.64/0.08/
```

```
***cycle***
Generation:   65  Best population fitness : 0.02428789 Mean of population:0.03580409
  Best chromosome->10.70/-19.86/36.83/-20.32/18.88/-39.12/-40.71/-0.97/-23.64/0.08/

***cycle***
Generation:   66  Best population fitness : 0.02428789 Mean of population:0.03565368
  Best chromosome->10.70/-19.86/36.83/-20.32/18.88/-39.12/-40.71/-0.97/-23.64/0.08/

***cycle***
Generation:   67  Best population fitness : 0.02428789 Mean of population:0.03563665
  Best chromosome->10.70/-19.86/36.83/-20.32/18.88/-39.12/-40.71/-0.97/-23.64/0.08/

***cycle***
Generation:   68  Best population fitness : 0.02428789 Mean of population:0.03554259
  Best chromosome->10.70/-19.86/36.83/-20.32/18.88/-39.12/-40.71/-0.97/-23.64/0.08/

***cycle***
Generation:   69  Best population fitness : 0.02428789 Mean of population:0.03540210
  Best chromosome->10.70/-19.86/36.83/-20.32/18.88/-39.12/-40.71/-0.97/-23.64/0.08/

***cycle***
Generation:   70  Best population fitness : 0.02428789 Mean of population:0.03533953
  Best chromosome->10.70/-19.86/36.83/-20.32/18.88/-39.12/-40.71/-0.97/-23.64/0.08/

***cycle***
Generation:   71  Best population fitness : 0.02428789 Mean of population:0.03521669
  Best chromosome->10.70/-19.86/36.83/-20.32/18.88/-39.12/-40.71/-0.97/-23.64/0.08/

***cycle***
Generation:   72  Best population fitness : 0.02428789 Mean of population:0.03497467
  Best chromosome->10.70/-19.86/36.83/-20.32/18.88/-39.12/-40.71/-0.97/-23.64/0.08/

***cycle***
Generation:   73  Best population fitness : 0.02428789 Mean of population:0.03496585
  Best chromosome->10.70/-19.86/36.83/-20.32/18.88/-39.12/-40.71/-0.97/-23.64/0.08/

***cycle***
Generation:   74  Best population fitness : 0.02428789 Mean of population:0.03482951
  Best chromosome->10.70/-19.86/36.83/-20.32/18.88/-39.12/-40.71/-0.97/-23.64/0.08/

***cycle***
Generation:   75  Best population fitness : 0.02428789 Mean of population:0.03478878
  Best chromosome->10.70/-19.86/36.83/-20.32/18.88/-39.12/-40.71/-0.97/-23.64/0.08/

***cycle***
Generation:   76  Best population fitness : 0.02428789 Mean of population:0.03472346
  Best chromosome->10.70/-19.86/36.83/-20.32/18.88/-39.12/-40.71/-0.97/-23.64/0.08/

***cycle***
```

Generation: 77 Best population fitness : 0.02301687 Mean of population:0.03458719
Best chromosome->27.32/-31.84/-3.00/-57.12/24.88/-16.14/-1.14/39.73/-1.15/0.18/

cycle
Generation: 78 Best population fitness : 0.02301687 Mean of population:0.03451069
Best chromosome->27.32/-31.84/-3.00/-57.12/24.88/-16.14/-1.14/39.73/-1.15/0.18/

cycle
Generation: 79 Best population fitness : 0.02301687 Mean of population:0.03434609
Best chromosome->27.32/-31.84/-3.00/-57.12/24.88/-16.14/-1.14/39.73/-1.15/0.18/

cycle
Generation: 80 Best population fitness : 0.02301687 Mean of population:0.03418806
Best chromosome->27.32/-31.84/-3.00/-57.12/24.88/-16.14/-1.14/39.73/-1.15/0.18/

cycle
Generation: 81 Best population fitness : 0.02301687 Mean of population:0.03402732
Best chromosome->27.32/-31.84/-3.00/-57.12/24.88/-16.14/-1.14/39.73/-1.15/0.18/

cycle
Generation: 82 Best population fitness : 0.02301687 Mean of population:0.03390054
Best chromosome->27.32/-31.84/-3.00/-57.12/24.88/-16.14/-1.14/39.73/-1.15/0.18/

cycle
Generation: 83 Best population fitness : 0.02301687 Mean of population:0.03387266
Best chromosome->27.32/-31.84/-3.00/-57.12/24.88/-16.14/-1.14/39.73/-1.15/0.18/

cycle
Generation: 84 Best population fitness : 0.02301687 Mean of population:0.03376543
Best chromosome->27.32/-31.84/-3.00/-57.12/24.88/-16.14/-1.14/39.73/-1.15/0.18/

cycle
Generation: 85 Best population fitness : 0.02301687 Mean of population:0.03370612
Best chromosome->27.32/-31.84/-3.00/-57.12/24.88/-16.14/-1.14/39.73/-1.15/0.18/

cycle
Generation: 86 Best population fitness : 0.02301687 Mean of population:0.03360450
Best chromosome->27.32/-31.84/-3.00/-57.12/24.88/-16.14/-1.14/39.73/-1.15/0.18/

cycle
Generation: 87 Best population fitness : 0.02301687 Mean of population:0.03355560
Best chromosome->27.32/-31.84/-3.00/-57.12/24.88/-16.14/-1.14/39.73/-1.15/0.18/

cycle
Generation: 88 Best population fitness : 0.02301687 Mean of population:0.03331063
Best chromosome->27.32/-31.84/-3.00/-57.12/24.88/-16.14/-1.14/39.73/-1.15/0.18/

cycle
Generation: 89 Best population fitness : 0.02301687 Mean of population:0.03308697

Best chromosome->27.32/-31.84/-3.00/-57.12/24.88/-16.14/-1.14/39.73/-1.15/0.18/

cycle
Generation: 90 Best population fitness : 0.02301687 Mean of population:0.03299376
 Best chromosome->27.32/-31.84/-3.00/-57.12/24.88/-16.14/-1.14/39.73/-1.15/0.18/

cycle
Generation: 91 Best population fitness : 0.02301687 Mean of population:0.03280551
 Best chromosome->27.32/-31.84/-3.00/-57.12/24.88/-16.14/-1.14/39.73/-1.15/0.18/

cycle
Generation: 92 Best population fitness : 0.02152323 Mean of population:0.03267332
 Best chromosome->27.32/-41.85/4.38/-14.27/24.88/-14.09/-9.88/39.73/-1.15/0.18/

cycle
Generation: 93 Best population fitness : 0.02152323 Mean of population:0.03265421
 Best chromosome->27.32/-41.85/4.38/-14.27/24.88/-14.09/-9.88/39.73/-1.15/0.18/

cycle
Generation: 94 Best population fitness : 0.02152323 Mean of population:0.03250503
 Best chromosome->27.32/-41.85/4.38/-14.27/24.88/-14.09/-9.88/39.73/-1.15/0.18/

cycle
Generation: 95 Best population fitness : 0.02152323 Mean of population:0.03202693
 Best chromosome->27.32/-41.85/4.38/-14.27/24.88/-14.09/-9.88/39.73/-1.15/0.18/

cycle
Generation: 96 Best population fitness : 0.02002237 Mean of population:0.03194529
 Best chromosome->0.29/-5.15/3.81/-11.03/17.01/-9.38/21.15/5.84/-0.76/0.17/

cycle
Generation: 97 Best population fitness : 0.02002237 Mean of population:0.03192516
 Best chromosome->0.29/-5.15/3.81/-11.03/17.01/-9.38/21.15/5.84/-0.76/0.17/

cycle
Generation: 98 Best population fitness : 0.02002237 Mean of population:0.03178475
 Best chromosome->0.29/-5.15/3.81/-11.03/17.01/-9.38/21.15/5.84/-0.76/0.17/

cycle
Generation: 99 Best population fitness : 0.02002237 Mean of population:0.03174589
 Best chromosome->0.29/-5.15/3.81/-11.03/17.01/-9.38/21.15/5.84/-0.76/0.17/

cycle
Generation: 100 Best population fitness : 0.02002237 Mean of population:0.03169513
 Best chromosome->0.29/-5.15/3.81/-11.03/17.01/-9.38/21.15/5.84/-0.76/0.17/

Call:
ANNGA.default(x = input, y = output, design = c(1, 3, 1), population = 100,
   mutation = 0.3, crossover = 0.7, maxGen = 100, error = 0.001)

```
***************************************************************
Mean Squared Error-------------------------------> 0.02002237
R2-----------------------------------------------> 0.4627776
Number of generation-----------------------------> 101
Weight range at initialization-------------------> [ 25 , -25 ]
Weight range resulted from the optimisation-----> [ 21.14924 , -11.02609 ]
***************************************************************
```

第 10 章 关联规则

关联规则最早是由 R.Agrawal 等人针对超市购物篮分析（Market Basket Analysis）问题而提出的，其目的是发现超市交易数据库中不同商品之间的关联关系。

本章重点内容：

- 关联规则简介
- Apriori算法

10.1 关联规则简介

关联规则呈现了顾客购物的行为模式，其结果可以作为经营决策、市场预测和制定销售策略的参考依据。以尿布和啤酒的购物篮为例：

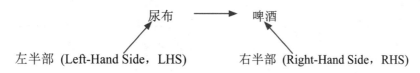

此规则表示尿布及啤酒销售具有关联性。

假设总共有 N 笔交易，定义 $supp(A)$ 为购买项目 A 的支持度（Support），$conf(A \rightarrow B)$ 为购买项目 A 也会购买项目 B 的置信度（Confidence）。支持度可用来判断规则的有效性；置信度用来判断在项目 A 的条件下发生项目 B 的可能性，其值越高，规则就越具有参考价值。一个强关联规则通常支持度和置信度值都高；但反过来，支持度和置信度值都高却不一定代表此关联规则所指的事件彼此间就一定存在高相关性，同时还需检查提升度（Lift）是否大于 1。提升度大于 1 表示项目 A 与项目 B 间有正向关系，提升度值等于 1 表示项目 A 与项目 B 间没有关系，提升度小于 1 表示项目 A 与项目 B 间为负向关系。支持度、置信度及提升度的计算方式如下：

$$supp(A) = freq(A)/N$$
$$supp(A \rightarrow B) = freq(A,B)/N$$
$$conf(A \rightarrow B) = freq(A,B)/freq(A)$$
$$lift(A \rightarrow B) = conf(A \rightarrow B)/supp(B)$$

其中：

- $freq(A)$ 表示购买项目 A 的数量。
- $freq(A,B)$ 表示同时购买项目 A 及 B 的数量。

10.2 Apriori 算法

1994 年，Agrawal & Srikant 提出了 Apriori 算法，是目前常用的关联规则。Apriori 算法采用一种逐层搜索的迭代方法（Level-Wise Search），先找出满足最小支持度的频繁项集（Frequent Itemsets，简称频集），再以最小置信度为条件计算频繁项集所形成的关联规则。当 Apriori 算法找出满足用户定义的最小支持度（Minimum Support）和最小置信度（Minimum Confidence）的关联规则时，这个规则才算成立。Apriori 算法步骤如下：

步骤01 找出频繁项集（Frequent Itemset）L_1。
重复步骤 2、3，直到无新频繁项集产生（k≥1）。

步骤02 获取长度为 $k+1$ 的候选项集（Candidate Itemset）C_{k+1}。
组合（Join）：将其中的项集两两组合为 C_{k+1}。
修剪（Prune）：修剪子集合不符合的候选项集 C_{k+1}，得到长度为 $k+1$ 的候选项集 C_{k+1}。

步骤03 找出长度为 k+1 的频繁项集 Lk+1。
计数（Count）：计算修剪后候选项集 Ck+1 的支持度。
删除（Delete）：删除支持度未达最小支持度的候选项集 Ck+1，产生长度为 k+1 的频繁项集。

步骤04 由频繁项集产生关联规则 L_{k+1}。

2000 年，Zaki 提出了 Eclat 算法，这是一种深度优先算法，具体做法是将交易数据库中的项（Item）作为键（Key），每项对应的交易编号（TID）作为值（Value）。Eclat 算法示意图如图 10-1 所示。

TID	Item
1	A, B
2	B, C
3	A, C
4	A, B, C

转换后

Item	TID
A	1, 3, 4
B	1, 2, 4
C	2, 3, 4

Item	Freq
A	3
B	3
C	3

计算长度为1的频繁项集

Item	Freq
A, B	2
A, C	2
B, C	2

由长度为1的频繁项集产生长度为2的频繁项集

Item	Freq
A, B, C	1

由长度为2的频繁项集产生长度为3的频繁项集

图 10-1 Eclat 算法

程序范例 10-1

首先使用 arules 程序包：

```
> library(arules)
```

使用 Adult 数据集：

```
> data("Adult")
```

设置支持度 = 0.5、置信度 = 0.9 以及不显示执行过程的相关信息（verbose=F）：

```
> rules <- apriori(Adult,parameter = list(supp = 0.5, conf = 0.9),control=list(verbose=F))
```

将 52 条关联规则按置信度排序并显示出来：

```
> rules.sorted=sort(rules,by="confidence")
> inspect(rules.sorted)
     lhs                              rhs                     support   confidence lift
1  {hours-per-week=Full-time}      => {capital-loss=None}   0.5606650 0.9582531 1.0052191
2  {workclass=Private}             => {capital-loss=None}   0.6639982 0.9564974 1.0033773
3  {workclass=Private,
    native-country=United-States}  => {capital-loss=None}   0.5897179 0.9554818 1.0023119
4  {capital-gain=None,
    hours-per-week=Full-time}      => {capital-loss=None}   0.5191638 0.9550659 1.0018756
5  {workclass=Private,
    race=White}                    => {capital-loss=None}   0.5674829 0.9549683 1.0017732
6  {workclass=Private,
    race=White,
    native-country=United-States}  => {capital-loss=None}   0.5181401 0.9535418 1.0002768
7  {}                              => {capital-loss=None}   0.9532779 0.9532779 1.0000000
8  {workclass=Private,
    capital-gain=None}             => {capital-loss=None}   0.6111748 0.9529145 0.9996188
9  {native-country=United-States}  => {capital-loss=None}   0.8548380 0.9525461 0.9992323
10 {workclass=Private,
    capital-gain=None,
    native-country=United-States}  => {capital-loss=None}   0.5414807 0.9517075 0.9983526
11 {race=White}                    => {capital-loss=None}   0.8136849 0.9516307 0.9982720
12 {workclass=Private,
    race=White,
    capital-gain=None}             => {capital-loss=None}   0.5204742 0.9511000 0.9977153
13 {race=White,
    native-country=United-States}  => {capital-loss=None}   0.7490480 0.9504325 0.9970152
14 {capital-gain=None}             => {capital-loss=None}   0.8706646 0.9490705 0.9955863
15 {capital-gain=None,
    native-country=United-States}  => {capital-loss=None}   0.7793702 0.9481891 0.9946618
16 {race=White,
    capital-gain=None}             => {capital-loss=None}   0.7404283 0.9470983 0.9935175
17 {sex=Male}                      => {capital-loss=None}   0.6331027 0.9470750 0.9934931
18 {sex=Male,
    native-country=United-States}  => {capital-loss=None}   0.5661316 0.9462068 0.9925823
19 {race=White,
```

```
     sex=Male}                         => {capital-loss=None}  0.5564268 0.9457804 0.9921350
20  {race=White,
     capital-gain=None,
     native-country=United-States} => {capital-loss=None}  0.6803980 0.9457029 0.9920537
21  {race=White, sex=Male,
     native-country=United-States} => {capital-loss=None}  0.5113632 0.9442722 0.9905529
22  {sex=Male,
     capital-gain=None}              => {capital-loss=None}  0.5696941 0.9415288 0.9876750
23  {sex=Male,
     capital-gain=None,
     native-country=United-States} => {capital-loss=None}  0.5084149 0.9404636 0.9865576
24  {hours-per-week=Full-time}    => {capital-gain=None}  0.5435895 0.9290688 1.0127342
25  {capital-loss=None,
     hours-per-week=Full-time}    => {capital-gain=None}  0.5191638 0.9259787 1.0093657
26  {workclass=Private}             => {capital-gain=None}  0.6413742 0.9239073 1.0071078
27  {workclass=Private,
     native-country=United-States} => {capital-gain=None}  0.5689570 0.9218444 1.0048592
28  {race=White}                     => {native-country=United-States} 0.7881127 0.9217231 1.0270761
29  {workclass=Private,
     race=White}                     => {capital-gain=None}  0.5472339 0.9208931 1.0038221
30  {race=White,
     capital-loss=None}             => {native-country=United-States} 0.7490480 0.9205626 1.0257830
31  {race=White,
     sex=Male}                        => {native-country=United-States} 0.5415421 0.9204803 1.0256912
32  {workclass=Private,
     capital-loss=None}             => {capital-gain=None}  0.6111748 0.9204465 1.0033354
33  {race=White,
     capital-gain=None}             => {native-country=United-States} 0.7194628 0.9202807 1.0254689
34  {race=White, sex=Male,
     capital-loss=None}             => {native-country=United-States} 0.5113632 0.9190124 1.0240556
35  {race=White, capital-gain=None,
     capital-loss=None}             => {native-country=United-States} 0.6803980 0.9189249 1.0239581
36  {workclass=Private, capital-loss=None,
     native-country=United-States} => {capital-gain=None}  0.5414807 0.9182030 1.0008898
37  {}                                 => {capital-gain=None}  0.9173867 0.9173867 1.0000000
38  {workclass=Private, race=White,
     capital-loss=None}             => {capital-gain=None}  0.5204742 0.9171628 0.9997559
39  {native-country=United-States} => {capital-gain=None}     0.8219565 0.9159062 0.9983862
40  {workclass=Private,
     race=White}                     => {native-country=United-States} 0.5433848 0.9144157 1.0189334
41  {race=White}                     => {capital-gain=None}  0.7817862 0.9143240 0.9966616
42  {capital-loss=None} => {capital-gain=None}  0.8706646 0.9133376 0.9955863
43  {workclass=Private, race=White,
     capital-loss=None}             => {native-country=United-States} 0.5181401 0.9130498 1.0174114
44  {race=White,
     native-country=United-States} => {capital-gain=None}  0.7194628 0.9128933 0.9951019
45  {capital-loss=None,
     native-country=United-States} => {capital-gain=None}  0.7793702 0.9117168 0.9938195
46  {race=White,
     capital-loss=None}             => {capital-gain=None}  0.7404283 0.9099693 0.9919147
```

```
47 {race=White, capital-loss=None,
    native-country=United-States} => {capital-gain=None} 0.6803980 0.9083504 0.9901500
48 {sex=Male}                     => {capital-gain=None} 0.6050735 0.9051455 0.9866565
49 {sex=Male,
    native-country=United-States} => {race=White}        0.5415421 0.9051090 1.0585540
50 {sex=Male,
    native-country=United-States} => {capital-gain=None} 0.5406003 0.9035349 0.9849008
51 {sex=Male,
    capital-loss=None,
    native-country=United-States} => {race=White}        0.5113632 0.9032585 1.0563898
52 {race=White,
    sex=Male}                     => {capital-gain=None} 0.5313050 0.9030799 0.9844048
```

设置支持度= 0.5、置信度= 0.9、不显示执行过程的相关信息以及不包含 race = White 和 sex = Male 项目集：

```
> is <- apriori(Adult, parameter = list(supp = 0.5, conf = 0.9),
+ appearance = list(none = c("race=White", "sex=Male")),
+ control=list(verbose=F))
```

确认无 race = White 和 sex = Male 项目集：

```
> itemFrequency(items(is))["race=White"]
race=White
         0
> itemFrequency(items(is))["sex=Male"]
sex=Male
       0
```

将 20 条关联规则按置信度排序并显示出来：

```
> is.sorted=sort(is,by="confidence")
> inspect(is.sorted)

    lhs                              rhs                    support   confidence lift
1   {hours-per-week=Full-time}    => {capital-loss=None} 0.5606650 0.9582531 1.0052191
2   {workclass=Private}           => {capital-loss=None} 0.6639982 0.9564974 1.0033773
3   {workclass=Private,
     native-country=United-States} => {capital-loss=None} 0.5897179 0.9554818 1.0023119
4   {capital-gain=None,
     hours-per-week=Full-time}    => {capital-loss=None} 0.5191638 0.9550659 1.0018756
5   {}                            => {capital-loss=None} 0.9532779 0.9532779 1.0000000
6   {workclass=Private,
     capital-gain=None}           => {capital-loss=None} 0.6111748 0.9529145 0.9996188
7   {native-country=United-States} => {capital-loss=None} 0.8548380 0.9525461 0.9992323
8   {workclass=Private,
     capital-gain=None,
     native-country=United-States} => {capital-loss=None} 0.5414807 0.9517075 0.9983526
9   {capital-gain=None}           => {capital-loss=None} 0.8706646 0.9490705 0.9955863
10  {capital-gain=None,
     native-country=United-States} => {capital-loss=None} 0.7793702 0.9481891 0.9946618
11  {hours-per-week=Full-time}    => {capital-gain=None} 0.5435895 0.9290688 1.0127342
```

```
12  {capital-loss=None,
     hours-per-week=Full-time}    => {capital-gain=None}  0.5191638  0.9259787  1.0093657
13  {workclass=Private}           => {capital-gain=None}  0.6413742  0.9239073  1.0071078
14  {workclass=Private,
     native-country=United-States} => {capital-gain=None}  0.5689570  0.9218444  1.0048592
15  {workclass=Private,
     capital-loss=None}           => {capital-gain=None}  0.6111748  0.9204465  1.0033354
16  {workclass=Private,
     capital-loss=None,
     native-country=United-States} => {capital-gain=None}  0.5414807  0.9182030  1.0008898
17  {}                            => {capital-gain=None}  0.9173867  0.9173867  1.0000000
18  {native-country=United-States} => {capital-gain=None}  0.8219565  0.9159062  0.9983862
19  {capital-loss=None}           => {capital-gain=None}  0.8706646  0.9133376  0.9955863
20  {capital-loss=None,
     native-country=United-States} => {capital-gain=None}  0.7793702  0.9117168  0.9938195
```

程序范例 10-2

首先使用 arules 程序包：

```
> library(arules)
```

使用 Titanic 数据集并转换为数据框对象 df：

```
> data("Titanic")
> str(Titanic)
 table [1:4, 1:2, 1:2, 1:2] 0 0 35 0 0 0 17 0 118 154 ...
 - attr(*, "dimnames")=List of 4
  ..$ Class   : chr [1:4] "1st" "2nd" "3rd" "Crew"
  ..$ Sex     : chr [1:2] "Male" "Female"
  ..$ Age     : chr [1:2] "Child" "Adult"
  ..$ Survived: chr [1:2] "No" "Yes"
> df <- as.data.frame(Titanic)
```

产生符合 apriori() 函数要求的新数据框对象 titanic.new：

```
> titanic.new <- NULL
> for(i in 1:4) {
+     titanic.new <- cbind(titanic.new, rep(as.character(df[,i]),df$Freq))
+ }
>
> titanic.new <- as.data.frame(titanic.new)
> names(titanic.new) <- names(df)[1:4]
> str(titanic.new)
'data.frame': 2201 obs. of  4 variables:
 $ Class   : Factor w/ 4 levels "1st","2nd","3rd",..: 3 3 3 3 3 3 3 3 3 3 ...
 $ Sex     : Factor w/ 2 levels "Female","Male": 2 2 2 2 2 2 2 2 2 2 ...
 $ Age     : Factor w/ 2 levels "Adult","Child": 2 2 2 2 2 2 2 2 2 2 ...
 $ Survived: Factor w/ 2 levels "No","Yes": 1 1 1 1 1 1 1 1 1 1 ...
```

在默认最小支持度 = 0.1、最小置信度 = 0.8 的条件下进行关联分析，获得 27 条规则：

```
> titanic_rules.all <- apriori(titanic.new)
```

```
Parameter specification:
 confidence minval smax arem  aval originalSupport support minlen maxlen target  ext
       0.8    0.1    1 none FALSE             TRUE     0.1      1     10  rules FALSE
Algorithmic control:
filter tree heap memopt load sort verbose
   0.1 TRUE TRUE  FALSE TRUE    2    TRUE

apriori - find association rules with the apriori algorithm
version 4.21 (2004.05.09) (c) 1996-2004 Christian Borgelt
set item appearances ...[0 item(s)] done [0.00s].
set transactions ...[10 item(s), 2201 transaction(s)] done [0.00s].
sorting and recoding items ... [9 item(s)] done [0.00s].
creating transaction tree ... done [0.00s].
checking subsets of size 1 2 3 4 done [0.00s]. writing ... [27 rule(s)] done [0.00s].
creating S4 object  ... done [0.00s].
> inspect(titanic_rules.all)
   lhs                rhs              support   confidence   lift
1  {}              => {Age=Adult}      0.9504771 0.9504771    1.0000000
2  {Class=2nd}     => {Age=Adult}      0.1185825 0.9157895    0.9635051
3  {Class=1st}     => {Age=Adult}      0.1449341 0.9815385    1.0326798
4  {Sex=Female}    => {Age=Adult}      0.1930940 0.9042553    0.9513700
5  {Class=3rd}     => {Age=Adult}      0.2848705 0.8881020    0.9343750
6  {Survived=Yes}  => {Age=Adult}      0.2971377 0.9198312    0.9677574
7  {Class=Crew}    => {Sex=Male}       0.3916402 0.9740113    1.2384742
8  {Class=Crew}    => {Age=Adult}      0.4020900 1.0000000    1.0521033
9  {Survived=No}   => {Sex=Male}       0.6197183 0.9154362    1.1639949
10 {Survived=No}   => {Age=Adult}      0.6533394 0.9651007    1.0153856
11 {Sex=Male}      => {Age=Adult}      0.7573830 0.9630272    1.0132040
12 {Sex=Female,
    Survived=Yes} => {Age=Adult}       0.1435711 0.9186047    0.9664669
13 {Class=3rd,
    Sex=Male}     => {Survived=No}     0.1917310 0.8274510    1.2222950
14 {Class=3rd,
    Survived=No}  => {Age=Adult}       0.2162653 0.9015152    0.9484870
15 {Class=3rd,
    Sex=Male}     => {Age=Adult}       0.2099046 0.9058824    0.9530818
16 {Sex=Male,
    Survived=Yes} => {Age=Adult}       0.1535666 0.9209809    0.9689670
17 {Class=Crew,
    Survived=No}  => {Sex=Male}        0.3044071 0.9955423    1.2658514
18 {Class=Crew,
    Survived=No}  => {Age=Adult}       0.3057701 1.0000000    1.0521033
19 {Class=Crew,
    Sex=Male}     => {Age=Adult}       0.3916402 1.0000000    1.0521033
20 {Class=Crew,
    Age=Adult}    => {Sex=Male}        0.3916402 0.9740113    1.2384742
21 {Sex=Male,
    Survived=No}  => {Age=Adult}       0.6038164 0.9743402    1.0251065
22 {Age=Adult,
    Survived=No}  => {Sex=Male}        0.6038164 0.9242003    1.1751385
```

```
23 {Class=3rd,
   Sex=Male, Survived=No} => {Age=Adult}    0.1758292 0.9170616 0.9648435
24 {Class=3rd,
   Age=Adult, Survived=No} => {Sex=Male} 0.1758292 0.8130252 1.0337773
25 {Class=3rd,
   Sex=Male, Age=Adult}   => {Survived=No} 0.1758292 0.8376623 1.2373791
26 {Class=Crew,
   Sex=Male, Survived=No} => {Age=Adult}    0.3044071 1.0000000 1.0521033
27 {Class=Crew,
   Age=Adult,
   Survived=No}           => {Sex=Male} 0.3044071 0.9955423 1.2658514
```

若对设定最小支持度 =0.005、最小置信度 =0.8 进行关联分析，可获得 12 条关联性规则并依照提升度来排序：

```
> rules <- apriori(titanic.new, control = list(verbose=F),
+ parameter = list(minlen=2, supp=0.005, conf=0.8),
+ appearance = list(rhs=c("Survived=No", "Survived=Yes"),
+ default="lhs"))
> quality(rules) <- round(quality(rules), digits=3)
> rules.sorted <- sort(rules, by="lift")
> inspect(rules.sorted)
   Lhs                  rhs           support confidence lift
1  {Class=2nd,
    Age=Child}        => {Survived=Yes} 0.011    1.000 3.096
2  {Class=2nd,
    Sex=Female,
    Age=Child}        => {Survived=Yes} 0.006    1.000 3.096
3  {Class=1st,
    Sex=Female}       => {Survived=Yes} 0.064    0.972 3.010
4  {Class=1st,
    Sex=Female,
    Age=Adult}        => {Survived=Yes} 0.064    0.972 3.010
5  {Class=2nd,
    Sex=Female}       => {Survived=Yes} 0.042    0.877 2.716
6  {Class=Crew,
    Sex=Female}       => {Survived=Yes} 0.009    0.870 2.692
7  {Class=Crew,
    Sex=Female,
    Age=Adult}        => {Survived=Yes} 0.009    0.870 2.692
8  {Class=2nd,
    Sex=Female,
    Age=Adult}        => {Survived=Yes} 0.036    0.860 2.663
9  {Class=2nd,
    Sex=Male,
    Age=Adult}        => {Survived=No}  0.070    0.917 1.354
10 {Class=2nd,
    Sex=Male}         => {Survived=No}  0.070    0.860 1.271
11 {Class=3rd,
    Sex=Male,
    Age=Adult}        => {Survived=No}  0.176    0.838 1.237
```

```
12 {Class=3rd,
    Sex=Male}        => {Survived=No}   0.192          0.827 1.222
```

在产生关联规则的结果中，若某些规则比其他规则提供的信息少或没有额外信息，则称这条规则为冗余（Redundancy）规则。一般而言，冗余规则的提升度与其相关的关联规则的提升度相同或者较低。从以下结果可知有 4 条冗余规则。

```
> subset.matrix <- is.subset(rules.sorted, rules.sorted)
> subset.matrix[lower.tri(subset.matrix, diag=T)] <- NA
> redundant <- colSums(subset.matrix, na.rm=T) >= 1
> which(redundant)
[1] 2 4 7 8
```

去除冗余规则后，得到 12 条关联规则：

```
> rules.pruned <- rules.sorted[!redundant]
> inspect(rules.pruned)
  lhs                rhs              support confidence lift
1 {Class=2nd,
   Age=Child}     => {Survived=Yes}   0.011   1.000      3.096
2 {Class=1st,
   Sex=Female}    => {Survived=Yes}   0.064   0.972      3.010
3 {Class=2nd,
   Sex=Female}    => {Survived=Yes}   0.042   0.877      2.716
4 {Class=Crew,
   Sex=Female}    => {Survived=Yes}   0.009   0.870      2.692
5 {Class=2nd,
   Sex=Male,
   Age=Adult}     => {Survived=No}    0.070   0.917      1.354
6 {Class=2nd,
   Sex=Male}      => {Survived=No}    0.070   0.860      1.271
7 {Class=3rd,
   Sex=Male,
   Age=Adult}     => {Survived=No}    0.176   0.838      1.237
8 {Class=3rd,
   Sex=Male}      => {Survived=No}    0.192   0.827      1.222
```

用户可对 lhs 的细节继续进行调整以进一步分析关联规则，本范例中对 lhs 的 "Class=1st"、"Class=2nd"、"Class=3rd" 及 "Age=Child"、"Age=Adult" 进一步分析并产生 6 条关联规则。

```
> rules <- apriori(titanic.new,
+ parameter = list(minlen=3, supp=0.002, conf=0.2),
+ appearance = list(rhs=c("Survived=Yes"),
+ lhs=c("Class=1st", "Class=2nd", "Class=3rd",
+ "Age=Child", "Age=Adult"),
+ default="none"),
+ control = list(verbose=F))
> rules.sorted <- sort(rules, by="confidence")
> inspect(rules.sorted)
  lhs                rhs              support     confidence lift
1 {Class=2nd,
   Age=Child}     => {Survived=Yes}   0.010904134 1.0000000  3.0956399
2 {Class=1st,
   Age=Child}     => {Survived=Yes}   0.002726034 1.0000000  3.0956399
3 {Class=1st,
   Age=Adult}     => {Survived=Yes}   0.089504771 0.6175549  1.9117275
4 {Class=2nd,
   Age=Adult}     => {Survived=Yes}   0.042707860 0.3601533  1.1149048
5 {Class=3rd,
```

```
    Age=Child}            => {Survived=Yes} 0.012267151 0.3417722 1.0580035
 6 {Class=3rd,
    Age=Adult}            => {Survived=Yes} 0.068605179 0.2408293 0.7455209
```

用户可再调用 arulesViz 程序包和 plot() 函数来产生关联规则散点图（见图 10-2）和 Two-key 图（见图 10-3）：

```
> library(arulesViz)
> plot(titanic_rules.all)
> plot(titanic_rules.all, shading="order", control=list(main =+"Two-key plot",col=rainbow(5)))
```

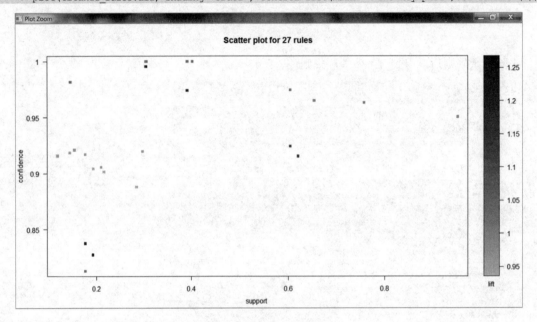

图 10-2　关联规则散点图

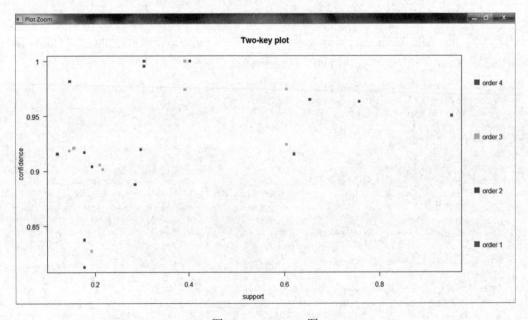

图 10-3　Two-key 图

第 11 章 社交网络分析和文本挖掘

在本章中，我们将介绍脸书（Facebook）的社交网络分析和文本挖掘。

本章重点内容：

- 社交网络分析
- 文本挖掘

11.1 社交网络分析

在本章中，用户可分析脸书的数据，首先到网址 https://developers.facebook.com/ 注册获取应用程序编号（App ID）、应用程序密钥（App Secret）或令牌（Token），用户可通过 R 程序来分析脸书的相关信息。App ID 和 App Secret 可长期使用，而令牌只可短期使用。

获取 App ID、App Secret 的步骤如下：

步骤01 先以用户账号登录到脸书，再连接到网址 https://developers.facebook.com/，如图 11-1 所示。

图 11-1 脸书的开发者网页

步骤02 选择"新建应用"选项，如图 11-2 所示。

图 11-2 新建应用的网页显示界面

步骤03 单击"网站"选项,网页显示将如图 11-3 所示。

图 11-3 脸书创建网站时的网页显示界面

步骤04 在网站 URL 字段中输入 http://localhost:1410/,并单击"Next"按钮,如图 11-4 所示。

社交网络分析和文本挖掘 第11章

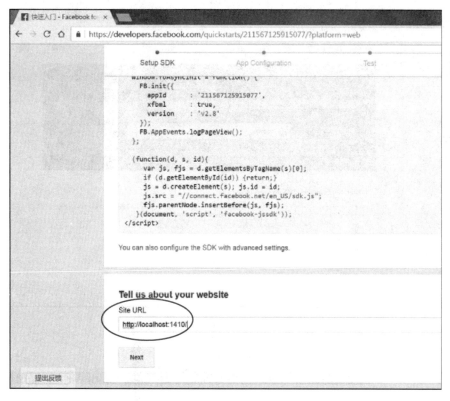

图 11-4 输入 URL 时的网页显示界面

步骤 05 单击最下方的"Finished!"按钮后,拖曳网页到最上方,再单击"Skip Quick Start"按钮,如图 11-5 和图 11-6 所示。

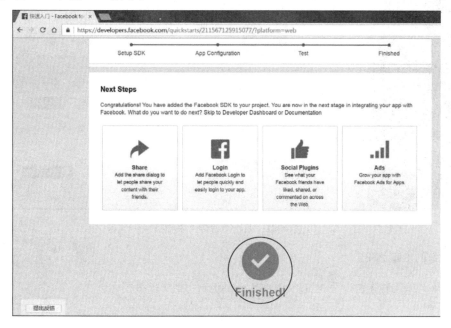

图 11-5 单击"Finished!"按钮

图 11-6 单击"Skip Quick Start"按钮

步骤 06 单击"显示"按钮,获取应用编号和应用密钥,如图 11-7 所示。

图 11-7 单击"显示"按钮

获取令牌的步骤如下:

步骤 01 选择"工具和支持"下的 Graph API Explorer,如图 11-8 所示。

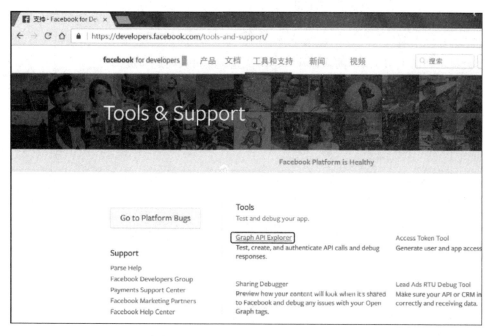

图 11-8 选择"Graph API Explorer"

步骤 02 获取令牌,如图 11-9 所示。

图 11-9 取得令牌

在本节中将示范应用编号（App ID）和应用密钥（App Secret）的使用。

程序范例 11-1

首先使用 Rfacebook 和 wordcloud 程序包，wordcloud 是文字云程序包：

```
> library(Rfacebook)
> library(wordcloud)
```

用户使用从脸书申请到的 App ID 和 App Secret，并调用 fbOAuth() 函数取得令牌（Token）：

```
> token <- fbOAuth(app_id="xx", app_secret="xx",extended_permissions
+                 = TRUE)

Copy and paste into Site URL on Facebook App Settings: http://localhost:1410/ When done, press
any key to continue...
Waiting for authentication in browser...
Press Esc/Ctrl + C to abort
```

```
Authentication complete.
Authentication successful.
```

注意会出现如图 11-10 所示的认证网页，确认无误后可关闭此网页。

图 11-10 认证网页

取得令牌后，调用 getUsers() 函数获取用户信息：

```
> me <- getUsers("me", token, private_info = TRUE)
```

确认名称无误：

```
> me$name # my name
[1] "JohnLee"
```

获取用户点赞的粉丝团，并调用 fix() 函数显示其信息：

```
> my_likes <- getLikes(user="me", token = token)
> fix(my_likes)
```

获取 2012/01/01 到 2014/12/22 期间粉丝团（page = "109249609124014"）的帖文信息（默认获取 100 篇帖文）：

```
> fb_page1<-getPage(page="109249609124014",token,
+             since='2012/01/01',until='2014/12/22')
100 posts
```

调用 wordcloud() 函数来显示帖文信息：

```
> wordcloud(fb_page1$message , fb_page1$comments_count)
```

调用 wordcloud() 函数来显示帖文点赞信息：

```
> wordcloud(fb_page1$message , fb_page1$likes_count)
```

11.2 文本挖掘

文本挖掘（Text Mining）与数据挖掘的不同之处在于要挖掘没有特定结构的纯文字，而这些文字内容中也可能蕴藏着有用的信息。

程序范例11-2

首先使用 gutenbergr、jiebaR、dplyr 和 wordcloud 程序包：

```
> library(gutenbergr)
```

```
> library(jiebaR)
> library(dplyr)
> library(wordcloud)
```

使用 jiebaR 程序包中 worker() 函数默认的混合分词类型:

```
> mixSeg <- worker()
```

下载 PG 网站 https://www.gutenberg.org/ 中编号为 27166 的中文电子书（鲁迅：呐喊）并赋值给 luxun 对象:

```
> luxun <- gutenberg_download(27166)
```

调用 segment() 函数对 luxun 对象的 text 变量分词并赋值给 luxun.seg 对象:

```
> str(luxun)
Classes 'tbl_df', 'tbl' and 'data.frame':  5798 obs. of  2 variables:
 $ gutenberg_id: int  27166 27166 27166 27166 27166 27166 27166 27166 27166 27166 ...
 $ text        : chr  "《鲁迅：呐喊》" "" "" "《呐喊》自序" ...
> luxun.seg <- segment(luxun$text, mixSeg)
```

调用 head() 函数显示前 6 项数据:

```
> luxun_head <- head(luxun.seg)
> luxun_head
[1] "鲁迅" "" ":" "" "呐喊" "呐喊" "自序" "呐喊"
```

调用 freq() 函数建立词频表并加上列名称:

```
> luxun.freq <- freq(luxun.seg)
> colnames(luxun.freq) <- c("word","freq")
```

调用 arrange() 函数制作一个排序过（从高到低）的频率表，显示前 6 项数据:

```
> freq_df <- arrange(luxun.freq, desc(freq))
> head(freq_df)
  word freq
1   的 2609
2   了 1423
3   他  825
4   是  606
5   我  599
6   也  597
```

使用 tag 重新分词、标注词性并赋值给 luxun.pos 对象:

```
> pos.tagger <- worker("tag")
> luxun.pos <- segment(luxun$text, pos.tagger)
```

调用 name() 函数获取词性后与 luxun.pos 对象建立数据框，加上列名称:

```
> tmp_df <- data.frame(luxun.pos, names(luxun.pos))
> colnames(tmp_df)<-c("Word","POS")
```

启用管道（Pipe）符号 %>% 功能，将 tmp_df 传送给 groupBy() 函数来分组化 Word 和 POS 列数据，再使用 %>% 传送到 summarize() 和 n() 函数来计算分组的总项数，最后将结果赋值给 Word_POS_Freq：

```
> tmp_df %>%
+    group_by(Word,POS) %>%
+    summarise(Frequency=n()) -> Word_POS_Freq
```

调用 arrange()函数制作一个排序过（从高到低）的词频表，显示前 6 项数据：

```
> Word_POS_Freq <- arrange(Word_POS_Freq, desc(Word,POS))
> head(Word_POS_Freq)
Source: local data frame [6 x 3]
Groups: Word [6]

  Word  POS Frequency
  <fctr> <fctr> <int>
1  齉    zg    1
2  颧骨   n     2
3  钻进   v     1
4  钻水   x     1
5  钻    v     1
6  锣声   n     1
```

使用 subset()函数取出 POS 为代名词（标记为r）的前 200 个词，算出各自的频率后用 wordcloud() 函数显示出来：

```
> pos_r <- head(subset(Word_POS_Freq,Word_POS_Freq$POS == "r"),200)
> pos_r
Source: local data frame [133 x 3]  Groups: Word [133]

    Word    POS Frequency
    <fctr>  <fctr>   <int>
1   该书     r        1
2   该是     r        2
3   该       r        9
4   这类     r        2
5   这边     r        2
6   这样     r        56
7   这么     r        13
8   这种     r        9
9   这话     r        7
10  这间     r        1
# ... with 123 more rows

> wordcloud(pos_r$Word,pos_r$Frequency)
```

第 12 章 图形化数据分析工具

图形化数据分析工具是让用户不需要使用太多 R 指令，直接使用图形化界面就可以达到数据分析的功能。本章将以 rattle 程序包来介绍。

本章重点内容：

- 导入数据
- 探索和检测数据
- 转换数据
- 建立、评估和导出模型

Rattle（The R Analytic Tool To Learn Easily）程序包是一个图形化的数据分析工具，允许用户从 CSV 文件、ARFF，使用 ODBC（Open Database Connectivity，开放数据库连接）、R Dataset、RData File、Library 或 Corpus 来导入数据、探索（Explore）和测试（Test）数据、转换（Transform）数据、建立（Build）及评估（Evaluate）模型（Model），并可导出模型来供用户自行修改和运用。

首先使用 rattle 程序包并执行 rattle()函数：

```
> library('rattle')
> rattle()
```

产生 rattle 执行窗口，如图 12-1 所示（注意此窗口与 RStudio 或 R 窗口不同，用户需自行切换窗口）。

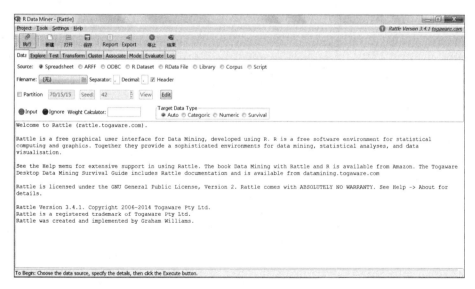

图 12-1 rattle 执行窗口

12.1 导入数据

用户可使用 rattle 来导入多种数据格式，包含 Spreadsheet、ARFF、ODBC、R Dataset、RData File、Library 及 Corpus。用户可使用 Spreadsheet 导入 CSV 和文本文件。若用户要导入 iris.csv 文件，则步骤为：① 选择 Data；② 选择 Source 中的 Spreadsheet；③ 下拉 Filename 菜单；④ 选择 CSV Files 并选择 iris.csv 文件；⑤ 单击执行按钮。操作顺序如图 12-2 所示。注意，可先按 2.2 节的方式产生 iris.csv 后再复制到 C:\下，并在 R 窗口执行以下指令来设置路径，再执行上述 5 个步骤。

```
> setwd("c:/")
> getwd()
[1] "c:/"
```

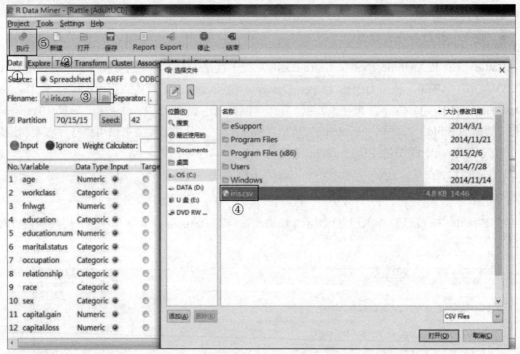

图 12-2　在 rattle 中导入 iris.csv

用户导入 CSV 文件时可设置分隔符（Separator）和 Decimal 符号，用户也可决定是否导入 Header。图 12-3 所示为导入 iris.csv 时设置 Separator 为 ","、Decimal 为 "." 以及要导入 Header。

图形化数据分析工具 第12章

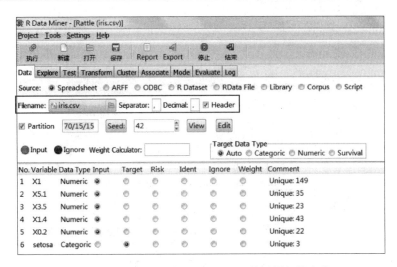

图 12-3　在 rattle 中导入 iris.csv 并设置相关参数

ARFF 文件是 Weka 默认的数据格式（arff），此格式是以文本文件格式保存的。若用户要导入 iris.arff 文件，则可参考导入 iris.csv 文件的方式（用户可到 http://tunedit.org/repo/UCI/iris.arff 下载 iris.arff 文件或使用本书提供的 iris.arff）。

用户要使用 ODBC 导入数据库数据时，可依次选择 Windows 10 下的"控制面板"→"系统和安全"→"管理工具"→"ODBC 数据源（64 位）"或"ODBC 数据源（32 位）"，并单击"添加"按钮来"创建新数据源"，具体步骤如图 12-4~图 12-7 所示。

图 12-4　单击"控制面板"中"系统和安全"下的"管理工具"

图 12-5　单击"ODBC 数据源"（64 位或者 32 位）

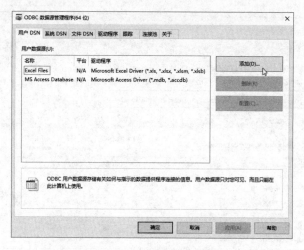

图 12-6　单击"添加"按钮

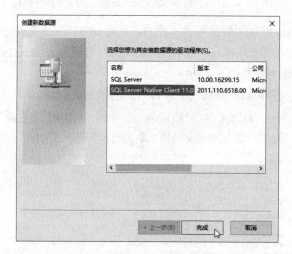

图 12-7　创建新数据源

用户要使用 ODBC 时，必须先到 R 窗口中加载 RODBC 程序包：

```
> library(RODBC)
```

创建用户数据源并加载 RODBC 程序包后，用户可在 Rattle 窗口中输入 DSN 名称（如 DSN:mitopac），再按 Enter 键来选择数据库表格（Table），最后单击"执行"按钮就可以导入数据，如图 12-8 所示。

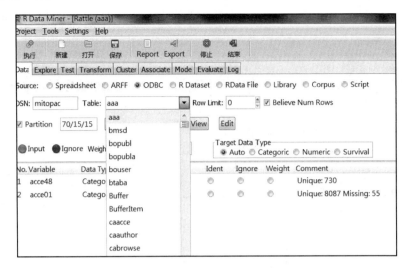

图 12-8　使用 ODBC 导入数据库数据

用户可使用 R Dataset 来导入数据（如 SPSS、SAS 及 DBF），使用 Corpus 导入语料库（Corpora）数据。用户可使用 RData File 导入 R 的 RData 格式文件，若用户要导入 iris.RData 文件，则步骤为① 选择 Data；② 选择 SourceR 中的 Data File；③ 下拉 Filename 菜单，选择 iris.RData 并设置 Data Name 为 iris；④ 单击"执行"按钮。具体操作步骤如图 12-9 所示。

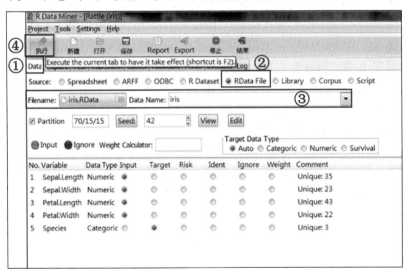

图 12-9　在 rattle 中导入 iris.RData

用户可使用 Library 来导入 R 程序中各个程序包的数据集，若用户想要使用 arules 的 Adult 数据集，则可先下拉 Data Name 菜单并选择 AdultUCI:arules:Adult Data Set 选项，再单击"执行"按钮。具体操作顺序如图 12-10 所示。

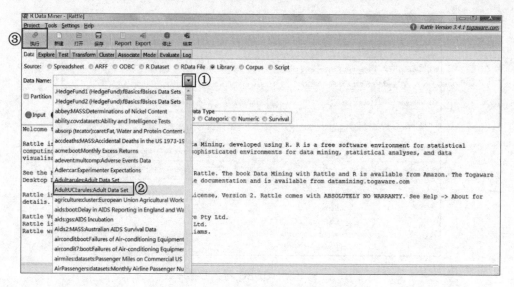

图 12-10　在 rattle 中导入 Adult 数据集

12.1.1　处理数据集

导入数据集后，Rattle 可再分割处理数据集为训练数据集（Training Dataset）、验证数据集（Validation Dataset）及测试数据集（Testing Dataset）。训练数据集用来建立模型，验证数据集用来调整模型参数以改进模型性能，而测试数据集用来真正评估模型的性能。用户可先选择 Partition 并改变训练数据集 / 验证数据集 / 测试数据集的比例（默认为 70/15/15），由于分割处理数据集采用的是随机方式，因此要固定数据集内容，可设置固定的 Seed 值，用户可进一步使用 View 查看数据集或使用 Edit 来修改数据集内的数据。图 12-11 所示为导入 iris.csv 后设置 Partition 为 80/10/10 以及设置 Seed 为 40 的操作界面。

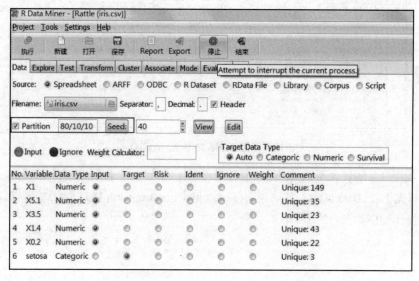

图 12-11　导入 iris.csv 后设置 Partition 和 Seed 的操作界面

12.1.2 设置变量

开始数据分析之前，必须了解 Rattle 提供不同类型变量（属性）的意义，Rattle 变量可以是模型的输入（Input）变量或构建模型的目标（Target）变量（输出变量）。变量还可以细分为风险（Risk）变量，即不建议在构建模型中使用此变量。忽略（Ignore）变量是构建模型时先暂时忽略不使用的变量；标识（Ident）变量是具有唯一性的标识符（例如日期和身份证）；权重（Weight）变量是可设置不同权值的变量，用户可使用 Weight Calculator 设置变量权重，例如 abs(X1) / max(X1)*10+1 表示设置变量 X1 的权重。大多数变量的默认作用都是输入变量（即独立变量、自变量）。目标变量用于构建模型的输出值（即依赖变量、应变量），变量的数据类型（Data Type）可以是数值（Numeric）类型或分类（Categoric）类型。Rattle 在导入数据时会默认将数值类型的变量设为目标变量，而分类类型的目标变量不适用于构建回归分析。用户可直接单击单选按钮（Radio Button）选择符合需要的变量及数据类型，例如设置 iris.csv 变量时，X1 变量的权重为 abs(X1)/max(X1)*10+1，X5.1、X3.5、X1.4 及 X0.2 为数值类型的输入变量，setosa 为分类类型的目标变量。图 12-12 所示为设置 iris.csv 变量的界面。

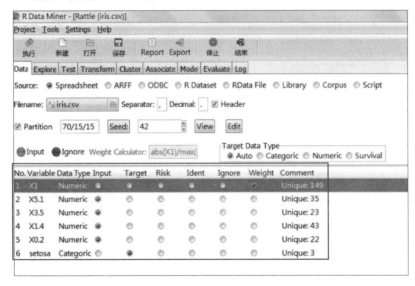

图 12-12　设置 iris.csv 变量的操作界面

12.2 探索和测试数据

导入数据后，可用 Rattle 进行摘要（Summary）分析、分布（Distributions）分析、相关（Correlation）分析和主成分（Principle Component）分析来探索和了解数据。数据的摘要分析可探索数据集中每个变量的信息。对于数值类型变量的摘要分析，可以包括最小值、最大值、中位数（Median）、平均值（Mean）、第一个四分位数和第三个四分位数（Quartile）；对于分类类型变量的摘要分析则提供了频率分布（Frequency Distributions）。以 iris.csv 为例，执行摘要分析的步骤为：① 选择 Explore；② 选择 Summary；③ 单击"执行"按钮。图 12-13 所示为 iris.csv 摘要分析的执行界面。分布分析提供各种图形来了解数据的分布信息，其图形包含盒形图（Box Plot）和直方图（Histogram）等。图 12-14 所示为设置 iris.csv 中 X1 变量执行盒形图分布分析，Rattle 的图形会显示在 RStudio 或 R 窗口中，如图 12-15 所示。相关分析提供了数值类型变量的相关性计算，并使用圆圈（Circle）和颜色来显示相关的强度，如图 12-16 和图 12-17 所示。主成分分析表示经由线性组合而得的主成分可以保有原来变量最多的信息，即主成分有最大的方差而且会显示出最大的个别差异，如图 12-18 和图 12-19 所示。

Rattle 也提供多项测试（Test）功能，其功能包含 Kolmogorov-Smirnov test、Wilcoxon test、T-test、F-test、Correlation test 及 Wilcoxon signed rank test，以 iris.csv 数据集的 X1 变量执行 T-test，其执行界面如图 12-20 所示。

图 12-13　iris.csv 摘要分析的执行界面

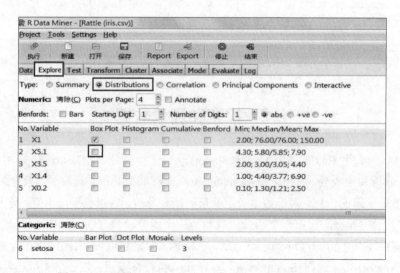

图 12-14　设置 iris.csv 中 X1 变量执行盒形图分布分析

图形化数据分析工具　**第 12 章**

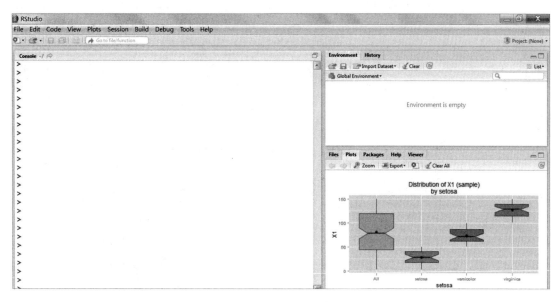

图 12-15　iris.csv 中 X1 变量的盒形图分布分析

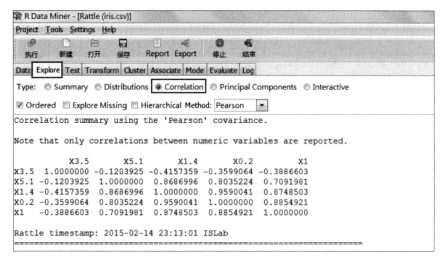

图 12-16　iris.csv 中各变量的相关分析值

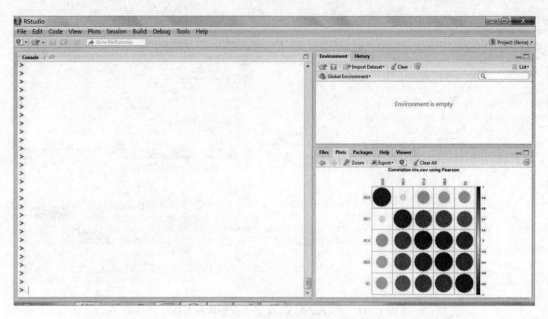

图 12-17 iris.csv 中各变量的相关分析图

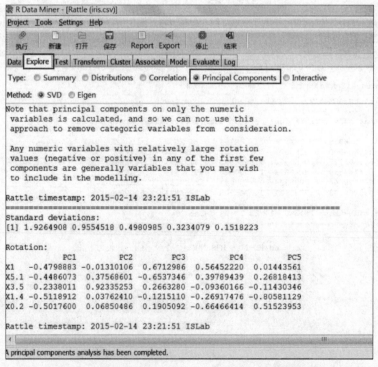

图 12-18 iris.csv 中各变量的主成分分析值

图形化数据分析工具　**第 12 章**

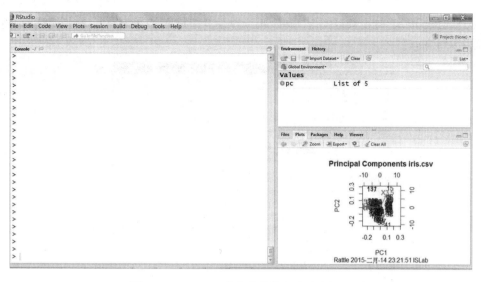

图 12-19　iris.csv 中各变量的主成分分析图

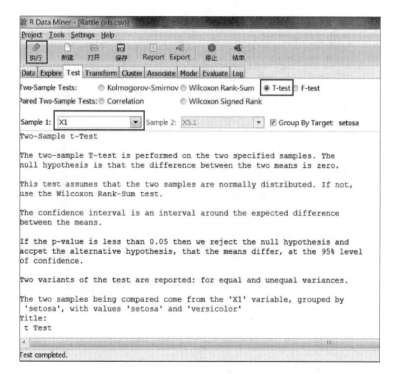

图 12-20　执行 iris.csv 数据集 X1 变量的 T-test 界面

12.3　转换数据

数据集中的数据可能存在缺失值，数据中若存在缺失值，则会严重影响数据分析的结果，甚至会影响建立模型的正确性。有些数据的变量并不存于数据集中，需通过转换而获得，所以必须对数据中的变量加以转换，以确保模型的质量。Rattle 中为转换数据提供了变量转换尺度（Rescale）、

· 135 ·

缺失值处理（Impute）、转换数据类型（Recode）以及清理数据（Cleanup）。转换数据的功能如图12-21~图12-24 所示。

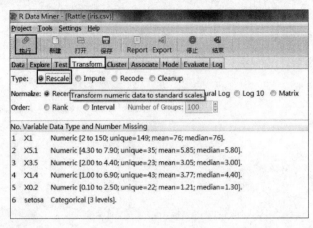

图 12-21　转换尺度

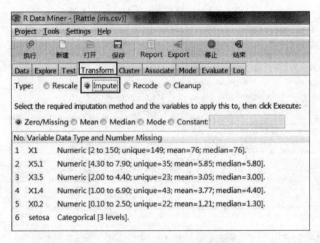

图 12-22　缺失值处理

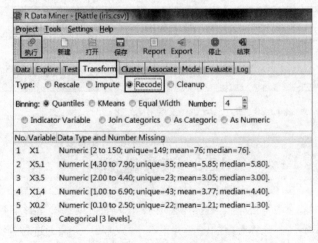

图 12-23　转换数据类型

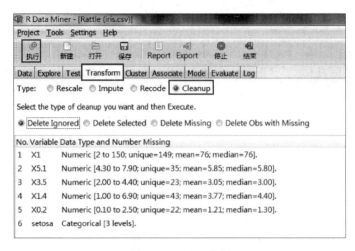

图 12-24　清理数据

12.4　建立、评估和导出模型

经过导入数据、探索数据和转换数据（这两者并非必要的步骤）后，用户就可以建立模型了。Rattle 提供聚类（Cluster）、关联规则（Associate）、分类及回归（Mode）等算法来建立模型。以 weather.csv 为例，用户导入数据后可建立聚类、关联规则、决策树及随机森林等模型，其执行界面如图 12-25~图 12-29 所示。

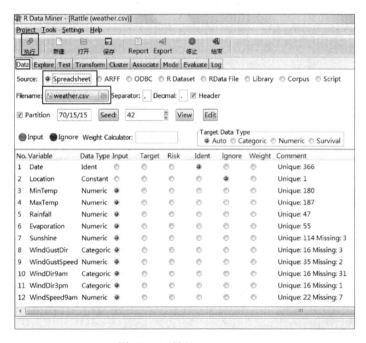

图 12-25　导入 weather.csv

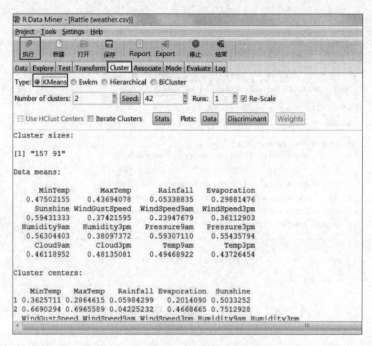

图 12-26 执行 K-Means 聚类算法

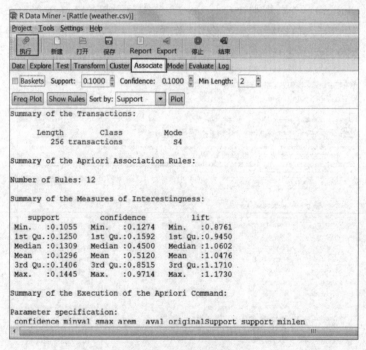

图 12-27 执行关联规则分析

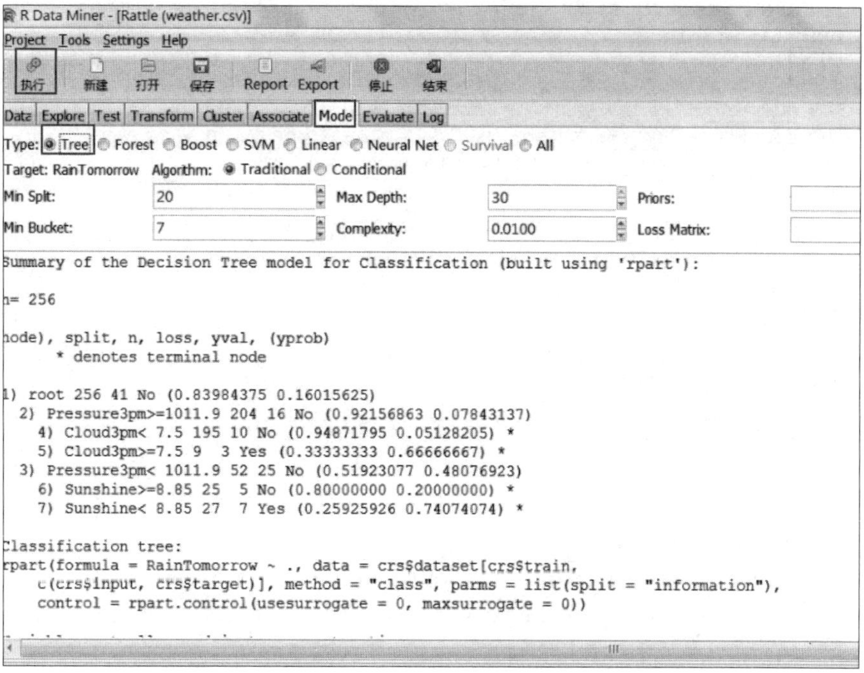

图 12-28　执行决策树算法

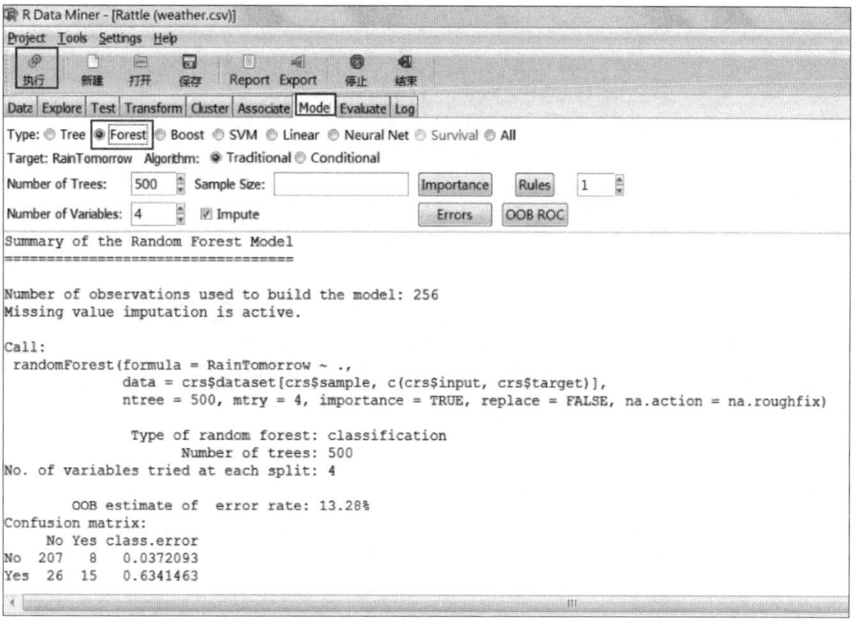

图 12-29　执行随机森林算法

建立模型后，用户可评估已经建立完成模型的性能。以 weather.csv 为例，用户可评估决策树和随机森林的误差矩阵（Error Matrix），其执行界面如图 12-30 所示。用户可使用 Rattle 提供的日志（Log）来查看程序代码，如图 12-31 所示。用户也可使用 pmml 程序包将 R 程序代码导出成 XML 格式的 PMML（Predictive Model Markup Language）。

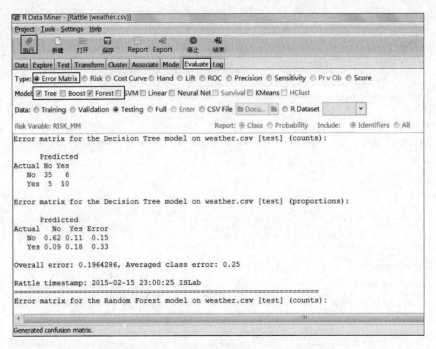

图 12-30　评估决策树及随机森林的误差矩阵

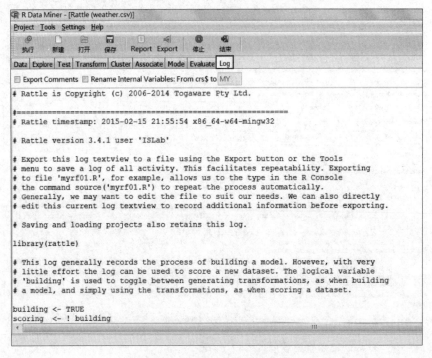

图 12-31　使用日志查看程序代码

第 13 章　大数据分析（R+Hadoop）

本章重点内容：
- Hadoop 简介
- R+Hadoop

13.1　Hadoop 简介

Hadoop 使用 MapReduce 分布式处理技术并使用 HDFS（Hadoop Distributed File System）作为分布式文件系统。MapReduce 是一种软件框架（Software Framework），此框架的主要功能是映射（Map）和化简（Reduce）。MapReduce 可在不同计算机组成的集群（Cluster）上执行，为大数据提供分布式计算和处理，可用 JAVA、R 或其他程序设计语言来实现其功能。Map 是从主节点（Master Node）输入一组 key/value 对（"键/值"对），将这组输入切分成好几个小的子部分，分散到各个"从"工作节点（Slave Nodes）进行计算或处理，输出则为另一组中间过程（Intermediate）的 key/value 对。Map 的功能如下：

$$(K_{in}, V_{in}) \rightarrow \text{list}(K_{inter}, V_{inter}) \tag{13-1}$$

Reduce 负责对相同中间过程的 key 合并其所有相关联的 Value，并产生输出结果的 key/value 对，将多对具有相同 Key 但不同 Value 的数据结合为多对的 Key/Value。Reduce 的功能如下：

$$(K_{inter}, \text{list}(V_{inter})) \rightarrow \text{list}(K_{out}, V_{out}) \tag{13-2}$$

目前应用 Hadoop 执行的任务已不仅是 MapReduce，在 Hadoop 2.x 版本中提供了 YARN（Yet Another Resource Negotiate）来处理分布在 Hadoop 集群的数据，YARN 可视为更通用的软件框架。

HDFS 在分布式存储环境中提供了单一的目录系统，数据以 Write-Once-Read-Many 方式存取。每个文件被分割成许多块（Block），每个块复制了许多复本（Replica），并分散存储于不同的数据节点（DataNode）上。HDFS 上的数据节点是实际存储文件块（Blocks）的服务器，数据节点负责维护 HDFS 的文件系统命名空间（File System Namespace）。

Hadoop 尚有许多外围项目组成 Hadoop 生态系统（Ecosystem），生态体系中的 HBase 是 Hadoop 上的数据库，它没有正规化与 Join 的概念，且利用 Family Columns 将相似的字段群聚在一起，以提高效率。Hive 是构建在 HDFS 上的一套分布式数据仓库系统，它提供的 HiveQL 语法让用户可以使用类似结构化的查询语言（Structured Query Language，SQL）来存取 Hadoop 文件中的数据集，例如 Join、Group by 等。Sqoop 是用来将 Hadoop 和关系数据库中的数据相互转移的工具，可以将一个关系数据库（例如 MS SQL）中的数据导入 Hadoop 的 HDFS 中，也可以将 HDFS 的数据导入关系数据库中。

13.2 R+Hadoop

R+Hadoop 结合 R 和 Hadoop 的功能，可扩大 R 处理数据的能力，让 R 可以进行分布式计算，使用 R 语言就可以轻易使用 Hadoop 功能。R+Hadoop 提供了用于 MapReduce、HDFS 及 HBase 的 rmr2、rhdfs 及 rhbase 程序包。rmr2 可以让用户开发并调用 MapReduce，rhdfs 让用户可以通过 R 存取 HDFS，rhbase 可以操作 HBase 数据。R+Hadoop 也提供了 RHive 程序包，RHive 可在 R 中便捷地使用 HiveQL 并加速分布式计算。

❖ rhdfs 常用指令如下：

启用 rhdfs	hdfs.init()
文件与目录的查看	hdfs.ls()
将文件从本地放到 HDFS 中	hdfs.put('test.txt', './')
复制文件	hdfs.copy('test.txt', 'test2.txt')
将文件下载到本地	hdfs.get('test.txt', '/home/test3.txt')
将文件移到不同位置	hdfs.move('test.txt', './test/q1.txt')
文件重新命名	hdfs.rename('./test/q1.txt', './test/test.txt')
删除文件	hdfs.rm('./test')
查看文件信息	hdfs.file.info('./')

❖ rhbase 常用指令如下：

hb.new.table	hb.delete.table	hb.describe.table
hb.set.table.mode	hb.regions.table	hb.insert
hb.get	hb.delete	hb.insert.data.frame
hb.get.data.frame	hb.scan	hb.list.tables
hb.defaults	hb.init	

❖ RHive 常用指令如下：

rhive.init	rhive.connect	rhive.set rhive.unset
rhive.query	rhive.execute rhive.big.query	rhive.assign
rhive.rm rhive.export	rhive.exportAll	rhive.list.udfs
rhive.rm.udf	rhive.script.export	rhive.script.unexport
rhive.close	rhive.list.databases	rhive.show.databases
rhive.use.database	rhive.list.tables	rhive.show.tables
rhive.desc.table	rhive.load.table	rhive.exist.table
rhive.size.table	rhive.drop.table	rhive.napply
rhive.sapply	rhive.aggregate	rhive.load
rhive.save	rhive.sample	rhive.mrapply
rhive.mapapply	rhive.reduceapply	rhive.hdfs.connect
rhive.hdfs.ls	rhive.hdfs.get	rhive.hdfs.put

rhive.hdfs.rm	rhive.hdfs.rename	rhive.hdfs.exists
rhive.hdfs.mkdirs	rhive.hdfs.cat	rhive.hdfs.tail
rhive.hdfs.du	rhive.hdfs.close	rhive.hdfs.info
rhive.hdfs.chmod	rhive.hdfs.chown	rhive.hdfs.chgrp
rhive.basic.mode	rhive.basic.range	rhive.basic.merge
rhive.basic.xtabs	rhive.basic.cut	rhive.basic.cut2
rhive.basic.by	rhive.basic.scale	rhive.basic.t.test
rhive.block.sample		

本节的范例并非完全在 Windows 环境下执行，用户执行本节范例前要先参考附录 E 来安装、设置及执行 R+Hadoop 虚拟机。在执行本章节范例时，需在 Windows 操作系统下使用浏览器（建议使用 Google Chrome 浏览器）输入 http://192.168.244.131:8787（本书提供单机版 R+Hadoop 的虚拟机网址，其设置方式可参考附录 E）连接到 RStudio，其执行界面如图 13-1 所示；接着输入 Username:hadoop 和 Password:hadoop 启动 RStudio，其执行界面如图 13-2 所示。

图 13-1　连接到 RStudio

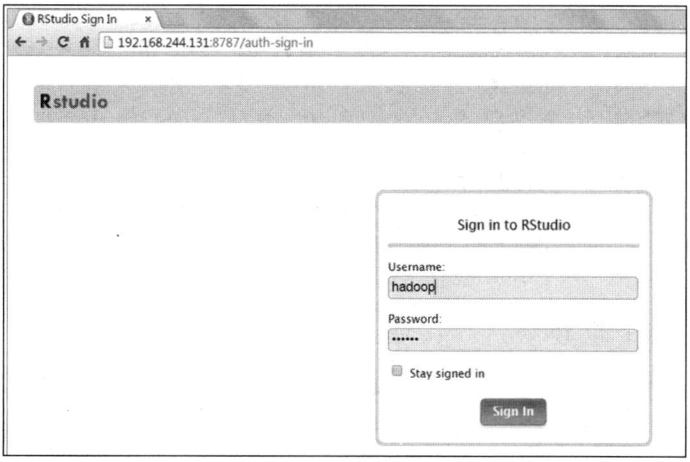

图 13-2　启动 RStudio

程序范例 13-1

先设置 R+Hadoop 的环境目录：

```
> Sys.setenv(
+    HIVE_HOME= '/usr/local/hive',
+    HADOOP_HOME= '/usr/local/hadoop-2.5.1',
+    HADOOP_CMD= '/usr/local/hadoop-2.5.1/bin/hadoop',
+    HADOOP_CONF_DIR= '/usr/local/hadoop-2.5.1/etc/hadoop',
+    HADOOP_STREAMING = '/usr/local/hadoop-2.5.1/share/hadoop/tools/lib/ hadoop-streaming-2.5.1.jar')
```

使用 rhdfs、rmr2、rhbase 和 RHive 程序包：

```
> library('rhdfs')
> library('rmr2')
> library('rhbase')
> library('RHive')
```

启用 hdfs：

```
> hdfs.init()
```

查看根目录下有 /tmp 目录：

```
> hdfs.ls("/")
  permission  owner    group size    modtime           file
1 drwxr-xr-x hadoop supergroup0 2015-03-11 07:29 /hbase
2 drwxr-xr-x hadoop supergroup0 2015-01-05 02:29 /rhive
3 drwxr-xr-x hadoop supergroup0 2015-01-05 02:32 /tmp
4 drwxr-xr-x hadoop supergroup0 2015-01-05 02:54 /user
```

查看 /tmp 目录下的文件：

```
> hdfs.ls("/tmp")
   permission     owner       group   size    modtime              file
1  -rw-r--r--     hadoop   supergroup  7493  2014-10-30  07:01  /tmp/file147e1f5316d3
2  drwxr-xr-x     hadoop   supergroup  0     2014-10-30  07:03  /tmp/file147e29e9e70
3  drwxr-xr-x     hadoop   supergroup  0     2015-01-05  02:26  /tmp/file156733f08c4a
4  -rw-r--r--     hadoop   supergroup  7457  2015-01-05  02:24  /tmp/file15676e6aabc2
5  -rw-r--r--     hadoop   supergroup  417   2015-01-05  02:19  /tmp/file1567790443ab
6  -rw-r--r--     hadoop   supergroup  7447  2014-11-01  11:32  /tmp/filed331eb5179e
7  drwxr-xr-x     hadoop   supergroup  0     2014-11-01  11:34  /tmp/filed334d03105f
8  -rw-r--r--     hadoop   supergroup  7498  2014-10-29  09:47  /tmp/filefb8640988a0
9  drwxr-xr-x     hadoop   supergroup  0     2014-10-29  09:49  /tmp/filefb86b5abff9
10 drwx------     hadoop   supergroup  0     2014-10-29  09:47  /tmp/hadoop-yarn
11 drwx-wx-wx     hadoop   supergroup  0     2015-01-04  20:54  /tmp/hive
```

产生临时文件 foo1 和 foo2：

```
> foo1 = tempfile()
> foo2 = tempfile()

> writeLines("foo1", con = foo1)
```

```
> writeLines("foo2", con = foo2)
```

文件从本地放到 HDFS 中：

```
> hdfs.put(foo1, "/tmp/foo1.txt")
[1] TRUE
> hdfs.put(foo2, "/tmp/foo2")
[1] TRUE
```

查看 /tmp 目录的 foo1.txt 和 foo2 文件：

```
> hdfs.ls("/tmp")
    permission     owner      group size            modtime                  file
1   -rw-r--r--    hadoop supergroup    5 2015-03-11 09:49 /tmp/file133a14a70a6f
2   -rwxrwxrwx    hadoop supergroup    5 2015-03-11 09:52 /tmp/file133a633650bd
3   -rwxrwxr--    hadoop supergroup    5 2015-03-11 09:52 /tmp/file133a7a15f55f
4   -rw-r--r--    hadoop supergroup 7493 2014-10-30 07:01 /tmp/file147e1f5316d3
5   drwxr-xr-x    hadoop supergroup    0 2014-10-30 07:03 /tmp/file147e29e9e70
6   drwxr-xr-x    hadoop supergroup    0 2015-01-05 02:26 /tmp/file156733f08c4a
7   -rw-r--r--    hadoop supergroup 7457 2015-01-05 02:24 /tmp/file15676e6aabc2
8   -rw-r--r--    hadoop supergroup  417 2015-01-05 02:19 /tmp/file1567790443ab
9   -rw-r--r--    hadoop supergroup 7447 2014-11-01 11:32 /tmp/filed331eb5179e
10  drwxr-xr-x    hadoop supergroup    0 2014-11-01 11:34 /tmp/filed334d03105f
11  -rw-r--r--    hadoop supergroup 7498 2014-10-29 09:47 /tmp/filefb8640988a0
12  drwxr-xr-x    hadoop supergroup    0 2014-10-29 09:49 /tmp/filefb86b5abff9
13  -rw-r--r--    hadoop supergroup    5 2015-03-11 11:07          /tmp/foo1.txt
14  -rw-r--r--    hadoop supergroup    5 2015-03-11 11:07              /tmp/foo2
15  drwx------    hadoop supergroup    0 2014-10-29 09:47       /tmp/hadoop-yarn
16  drwx-wx-wx    hadoop supergroup    0 2015-01-04 20:54              /tmp/hive
```

删除文件 foo1.txt：

```
> hdfs.delete("/tmp/foo1.txt")
15/03/11 11:07:39 INFO fs.TrashPolicyDefault: Namenode trash configuration:
Deletion interval = 0 minutes, Emptier interval = 0 minutes.
Deleted hdfs://master:9000/tmp/foo1.txt
[1] TRUE
```

查看 /tmp 目录下的文件，文件 foo1.txt 确实已删除：

```
> hdfs.ls("/tmp")
    permission     owner      group size            modtime                  file
1   -rw-r--r--    hadoop supergroup    5 2015-03-11 09:49 /tmp/file133a14a70a6f
2   -rwxrwxrwx    hadoop supergroup    5 2015-03-11 09:52 /tmp/file133a633650bd
3   -rwxrwxr--    hadoop supergroup    5 2015-03-11 09:52 /tmp/file133a7a15f55f
4   -rw-r--r--    hadoop supergroup 7493 2014-10-30 07:01 /tmp/file147e1f5316d3
5   drwxr-xr-x    hadoop supergroup    0 2014-10-30 07:03 /tmp/file147e29e9e70
6   drwxr-xr-x    hadoop supergroup    0 2015-01-05 02:26 /tmp/file156733f08c4a
7   -rw-r--r--    hadoop supergroup 7457 2015-01-05 02:24 /tmp/file15676e6aabc2
8   -rw-r--r--    hadoop supergroup  417 2015-01-05 02:19 /tmp/file1567790443ab
9   -rw-r--r--    hadoop supergroup 7447 2014-11-01 11:32 /tmp/filed331eb5179e
10  drwxr-xr-x    hadoop supergroup    0 2014-11-01 11:34 /tmp/filed334d03105f
11  -rw-r--r--    hadoop supergroup 7498 2014-10-29 09:47 /tmp/filefb8640988a0
```

```
12 drwxr-xr-x hadoop supergroup    0  2014-10-29  09:49  /tmp/filefb86b5abff9
13 -rw-r--r-- hadoop supergroup    5  2015-03-11  11:07  /tmp/foo2
14 drwx------ hadoop supergroup    0  2014-10-29  09:47  /tmp/hadoop-yarn
15 drwx-wx-wx hadoop supergroup    0  2015-01-04  20:54  /tmp/hive
```

启用 hbase：

```
> hb.init()
<pointer: 0x5b6dbc0>
attr(,"class")
[1] "hb.client.connection"
```

查看数据表：

```
> hb.list.tables()
list()
```

新建 test 数据表：

```
> hb.new.table('test','x','y','z')
[1] TRUE
```

增加 4 项数据：

```
> hb.insert("test",list(list("20100101",c("x:a","x:f","y","y:w"), list("James Dewey",TRUE, 187.5,189000))))
[1] TRUE
> hb.insert("test",list(list("20100102",c("x:a"), list("James Agnew"))))
[1] TRUE
> hb.insert("test",list(list("20100103",c("y:a","y:w"), list("Dilbert Ashford",250000))))
[1] TRUE
> hb.insert("test",list(list("20100104",c("x:f"), list("Henry Higs"))))
[1] TRUE
```

读取数据：

```
> hb.get("test",list("20100101","20100102")) [[1]]
[[1]][[1]]
[1] "20100101"

[[1]][[2]]
[1] "x:a" "x:f" "y:w"

[[1]][[3]]
[[1]][[3]][[1]]
[1] "James Dewey"

[[1]][[3]][[2]]
[1] TRUE

[[1]][[3]][[3]]
[1] 189000
```

```
[[2]]
[[2]][[1]]
[1] "20100102"

[[2]][[2]]
[1] "x:a"

[[2]][[3]]
[[2]][[3]][[1]]
[1] "James Agnew"
```

读取数据表中列字段开头是 y 的数据：

```
> hb.get("test",list("20100101","20100102"),c("y")) [[1]]
[[1]][[1]]
[1] "20100101"

[[1]][[2]]
[1] "y:w"

[[1]][[3]]
[[1]][[3]][[1]]
[1] 189000

[[2]]
[[2]][[1]]
[1] "20100102"

[[2]][[2]] NULL

[[2]][[3]]
list()
```

删除数据表数据：

```
> hb.delete("test","20100103","y:a")
[1] TRUE
```

确认数据已删除：

```
> hb.get("test","20100103")
[[1]]
[[1]][[1]]
[1] "20100103"

[[1]][[2]]
[1] "y:w"

[[1]][[3]]
[[1]][[3]][[1]]
[1] 250000
```

删除 test 数据表：

```
> hb.delete.table("test")
[1] TRUE
```

程序范例 13-2

先设置 R+Hadoop 的环境目录：

```
> Sys.setenv(
+    HIVE_HOME= '/usr/local/hive',
+    HADOOP_HOME='/usr/local/hadoop-2.5.1',
+    HADOOP_CMD='/usr/local/hadoop-2.5.1/bin/hadoop',
+    HADOOP_CONF_DIR='/usr/local/hadoop-2.5.1/etc/hadoop',
+    HADOOP_STREAMING           ='/usr/local/hadoop-2.5.1/share/hadoop/tools/lib/hadoop-streaming-2.5.1.jar')
```

使用 rhdfs、rmr2、rhbase 及 RHive 程序包：

```
> ## loading the libraries
> library('rhdfs')
> library('rmr2')
> library('rhbase')
> library('RHive')
```

启用 hdfs：

```
> hdfs.init()
```

产生 20 个二项分布的随机数，并将数据存储到 hdfs：

```
> groups <-to.dfs(keyval(NULL, rbinom(20, n=10, prob=0.4)))
15/03/12 20:44:24 INFO zlib.ZlibFactory: Successfully loaded & initialized native-zlib library
15/03/12 20:44:24 INFO compress.CodecPool: Got brand-new compressor [.deflate]
```

使用 mapreduce() 函数，其中 input 表示 group 的值为其 value，map 的作用是产生中间过程的 key/value 对，并将各 value 计数为 1，reduce 负责对相同中间过程的 key 加总，即所有 value 值的总和，最后产生结果的 key/value 对。

```
> out <-mapreduce(
+ input=groups,
+ map=function(k, v) keyval(v,1),
+ reduce=function(k, v) keyval(k, sum(v))
+ )
packageJobJar: [/usr/local/hadoop-2.5.1/tmp/hadoop-unjar6032394475589660817/]
[] /tmp/streamjob2635483612111907663.jar tmpDir=null
15/03/12 20:44:35 INFO client.RMProxy: Connecting to ResourceManager at /0.0.0.0:8032
15/03/12 20:44:36 INFO client.RMProxy: Connecting to ResourceManager at /0.0.0.0:8032
15/03/12 20:44:37 INFO mapred.FileInputFormat: Total input paths to process : 1
15/03/12 20:44:38 INFO mapreduce.JobSubmitter: number of splits:2
```

```
15/03/12 20:44:38 INFO mapreduce.JobSubmitter: Submitting tokens for job: job_1426206027775_0002
15/03/12 20:44:38 INFO impl.YarnClientImpl: Submitted application application_1426206027775_0002
15/03/12 20:44:38 INFO mapreduce.Job: The url to track the job: http://master:8088/proxy/application_1426206027775_0002/
15/03/12 20:44:38 INFO mapreduce.Job: Running job: job_1426206027775_0002
15/03/12 20:44:44 INFO mapreduce.Job: Job job_1426206027775_0002 running in uber mode : false
15/03/12 20:44:44 INFO mapreduce.Job:  map 0% reduce 0%
15/03/12 20:44:54 INFO mapreduce.Job:  map 50% reduce 0%
15/03/12 20:45:01 INFO mapreduce.Job:  map 100% reduce 0%
15/03/12 20:45:07 INFO mapreduce.Job:  map 100% reduce 100%
15/03/12 20:45:07 INFO mapreduce.Job: Job job_1426206027775_0002 completed successfully
15/03/12 20:45:07 INFO mapreduce.Job: Counters: 50
    File System Counters
        FILE: Number of bytes read=1068
        FILE: Number of bytes written=309257
        FILE: Number of read operations=0
        FILE: Number of large read operations=0
        FILE: Number of write operations=0
        HDFS: Number of bytes read=970
        HDFS: Number of bytes written=1354
        HDFS: Number of read operations=13
        HDFS: Number of large read operations=0
        HDFS: Number of write operations=2
    Job Counters
        Launched map tasks=2 Launched reduce tasks=1
        Data-local map tasks=2
        Total time spent by all maps in occupied slots (ms)=46936
        Total time spent by all reduces in occupied slots (ms)=19156
        Total time spent by all map tasks (ms)=11734
        Total time spent by all reduce tasks (ms)=4789
        Total vcore-seconds taken by all map tasks=11734
        Total vcore-seconds taken by all reduce tasks=4789
        Total megabyte-seconds taken by all map tasks=48062464
        Total megabyte-seconds taken by all reduce tasks=19615744
    Map-Reduce Framework
    Map input records=3
        Map output records=13
        Map output bytes=1036
        Map output materialized bytes=1074
        Input split bytes=182
        Combine input records=0
        Combine output records=0
        Reduce input groups=6
        Reduce shuffle bytes=1074
        Reduce input records=13
        Reduce output records=14
        Spilled Records=26
        Shuffled Maps =2
        Failed Shuffles=0
        Merged Map outputs=2
```

```
        GC time elapsed (ms)=245
        CPU time spent (ms)=5220
        Physical memory (bytes) snapshot=715452416
     Virtual memory (bytes) snapshot=3048878080
     Total committed heap usage (bytes)=531628032
   Shuffle Errors
        BAD_ID=0
        CONNECTION=0
        IO_ERROR=0
        WRONG_LENGTH=0
        WRONG_MAP=0
        WRONG_REDUCE=0
   File Input Format Counters
        Bytes Read=788
   File Output Format Counters
        Bytes Written=1354
   rmr
        reduce calls=6
15/03/12 20:45:07 INFO streaming.StreamJob: Output directory: /tmp/
file129a45af9b7c
```

将计算结果的 key/value 转换为数据框对象并显示前 10 项数据：

```
> outDF <- as.data.frame(from.dfs(out))
> head(outDF, 10)
  key val
1   5   1
2   7   2
3   8   2
4   9   1
5  10   3
6  11   1
```

绘出结果的数据框对象（见图 13-3）：

```
> with(outDF, barplot(val, names.arg=key))
```

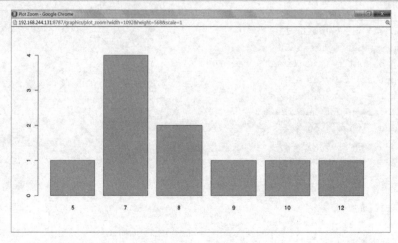

图 13-3　结果图

程序范例 13-3

先设置 R+Hadoop 的环境目录：

```
> Sys.setenv(
+    HIVE_HOME= '/usr/local/hive',
+    HADOOP_HOME= '/usr/local/hadoop-2.5.1',
+    HADOOP_CMD= '/usr/local/hadoop-2.5.1/bin/hadoop',
+    HADOOP_CONF_DIR= '/usr/local/hadoop-2.5.1/etc/hadoop',
+    HADOOP_STREAMING       =       '/usr/local/hadoop-2.5.1/share/hadoop/tools/lib/hadoop-streaming-2.5.1.jar')
```

使用 rhdfs、rmr2、rhbase 及 RHive 程序包：

```
> library('rhdfs')
> library('rmr2')
> library('rhbase')
> library('RHive')
```

启用 hdfs 并设置 rmr.options() 函数在本地执行：

```
> hdfs.init()
> rmr.options(backend = 'local')
NULL
```

注意，因为设置为在本地执行，所以没有程序范例 13-2 那样的 mapreduce 执行过程：

```
> groups <-to.dfs(keyval(NULL, rbinom(20, n=10, prob=0.5)))
> out <-mapreduce(
+ input=groups,
+ map=function(k, v) keyval(v,1),
+ reduce=function(k, v) keyval(k, sum(v))
+ )
```

将所得结果的 key/value 转换为数据框对象并显示前 10 项数据：

```
> outDF <- as.data.frame(from.dfs(out))
> head(outDF, 10)
  key val
1  10   2
2  11   4
3   9   2
4   4   1
5  13   1
```

程序范例 13-4

先设置 R+Hadoop 的环境目录：

```
> Sys.setenv(
+    HIVE_HOME= '/usr/local/hive',
+    HADOOP_HOME='/usr/local/hadoop-2.5.1',
```

```
+    HADOOP_CMD='/usr/local/hadoop-2.5.1/bin/hadoop',
+    HADOOP_CONF_DIR='/usr/local/hadoop-2.5.1/etc/hadoop',
+    HADOOP_STREAMING ='/usr/local/hadoop-2.5.1/share/hadoop/tools/lib/hadoop-streaming-2.5.1.jar')
```

使用 rhdfs、rmr2、rhbase、RHive 程序包并启用 hdfs：

```
> library('rhdfs')
Loading required package: rJava

HADOOP_CMD=/usr/local/hadoop-2.5.1/bin/hadoop

Be sure to run hdfs.init()
> library('rmr2')
> library('rhbase')
> library('RHive')
> hdfs.init()
15/03/12 23:30:02 WARN util.NativeCodeLoader: Unable to load native-hadoop
library for your platform... using builtin-java classes where applicable
```

选择以 hadoop 模式执行程序并使用 iris 数据集：

```
> #rmr.options(backend = 'local')
> rmr.options(backend = 'hadoop')
NULL
> data(iris)
```

用户自行定义 mapreduce 的 kmeans 函数（可参考 Vignesh Prajapati 编写的《Big Data Analytics with R and Hadoop》一书，Packt 出版）。

```
> mapreduce_kmeans =
+    function(
+      input_data,
+      cluster_number,
+      iter_number,
+      combine) {
+      ## Sum of square distant function
+      dist.fun =
+        function(C, input_data) {
+          apply(
+            C,
+            1,
+            function(x)
+              colSums((t(input_data) - x)^2))}
+
+      ## Map job
+      kmeans.map =
+        function(.,input_data) {
+          nearest = {
+            if(is.null(C))
+              sample(
+                1:cluster_number,
```

```
+             nrow(input_data),
+             replace = TRUE)
+      else {
+        D = dist.fun(C, input_data)
+        nearest = max.col(-D)}}
+    if(!combine)
+      keyval(nearest, input_data)
+    else
+      keyval(nearest, cbind(1,input_data))}
+  ## Reduce job
+  kmeans.reduce = {
+    if (!(combine) )
+      function(.,input_data)
+        t(as.matrix(apply(input_data, 2, mean)))
+    else
+      function(k, input_data)
+        keyval(
+          k,
+          t(as.matrix(apply(input_data, 2, sum))))}
+
+  C = NULL
+  for(i in 1:iter_number ) {
+    C =
+      values(
+
+        from.dfs(
+          mapreduce(
+            input_data,
+            map = kmeans.map,
+            reduce = kmeans.reduce)))
+    if(combine)
+      C = C[, -1]/C[, 1]
+
+    if(nrow(C) < cluster_number) {
+      C =
+        rbind(
+          C,
+          matrix(
+            rnorm(
+              (cluster_number -
+                nrow(C)) * nrow(C)),
+              ncol = nrow(C)) %*% C ) }}
+  C}
```

使用 iris 数据集执行 kmeans 并设置迭代次数=5：

```
> output= mapreduce_kmeans(
+    input_data=to.dfs(iris[,1:4]),
+    cluster_number = 3,
+    iter_number = 5,
+    combine = FALSE)
```

```
    15/03/12 23:30:21 INFO zlib.ZlibFactory: Successfully loaded & initialized native-zlib library
    15/03/12 23:30:21 INFO compress.CodecPool: Got brand-new compressor [.deflate] packageJobJar:
[/usr/local/hadoop-2.5.1/tmp/hadoop-unjar4591052676923830193/] []
/tmp/streamjob6349821224899197613.jar tmpDir=null
    15/03/12 23:30:24 INFO client.RMProxy: Connecting to ResourceManager at/0.0.0.0:8032
    15/03/12 23:30:24 INFO client.RMProxy: Connecting to ResourceManager at/0.0.0.0:8032
    15/03/12 23:30:25 INFO mapred.FileInputFormat: Total input paths to process : 1 15/03/12
23:30:25 INFO mapreduce.JobSubmitter: number of splits:2
    15/03/12 23:30:25 INFO mapreduce.JobSubmitter: Submitting tokens for job:
job_1426206027775_0023
    15/03/12 23:30:25 INFO impl.YarnClientImpl: Submitted application
application_1426206027775_0023
    15/03/12 23:30:26 INFO mapreduce.Job: The url to track the job: http://
master:8088/proxy/application_1426206027775_0023/
    15/03/12 23:30:26 INFO mapreduce.Job: Running job: job_1426206027775_0023 15/03/12
23:30:32 INFO mapreduce.Job: Job job_1426206027775_0023 running in uber mode : false
    15/03/12 23:30:32 INFO mapreduce.Job:  map 0% reduce 0%
    15/03/12 23:30:39 INFO mapreduce.Job:  map 50% reduce 0%
    15/03/12 23:30:46 INFO mapreduce.Job:  map 100% reduce 0%
    15/03/12 23:30:51 INFO mapreduce.Job:  map 100% reduce 100%
    15/03/12 23:30:52 INFO mapreduce.Job: Job job_1426206027775_0023 completed successfully
    15/03/12 23:30:53 INFO mapreduce.Job: Counters: 50
        File System Counters
            FILE: Number of bytes read=7294
            FILE: Number of bytes written=321724
            FILE: Number of read operations=0
            FILE: Number of large read operations=0
            FILE: Number of write operations=0
            HDFS: Number of bytes read=3018
            HDFS: Number of bytes written=1578
            HDFS: Number of read operations=13
            HDFS: Number of large read operations=0
            HDFS: Number of write operations=2
        Job Counters
            Launched map tasks=2
            Launched reduce tasks=1
            Data-local map tasks=2
            Total time spent by all maps in occupied slots (ms)=39600
            Total time spent by all reduces in occupied slots (ms)=19012
            Total time spent by all map tasks (ms)=9900
            Total time spent by all reduce tasks (ms)=4753
            Total vcore-seconds taken by all map tasks=9900
            Total vcore-seconds taken by all reduce tasks=4753
            Total megabyte-seconds taken by all map tasks=40550400
            Total megabyte-seconds taken by all reduce tasks=19468288
        Map-Reduce Framework
            Map input records=3
            Map output records=7
            Map output bytes=7260
```

```
        Map output materialized bytes=7300
        Input split bytes=182
        Combine input records=0
        Combine output records=0
        Reduce input groups=3
        Reduce shuffle bytes=7300
        Reduce input records=7
        Reduce output records=6
        Spilled Records=14
        Shuffled Maps =2
        Failed Shuffles=0
        Merged   Map outputs=2
        GC time elapsed (ms)=112
        CPU time spent (ms)=3020
        Physical memory (bytes) snapshot=697237504
        Virtual memory (bytes) snapshot=3079880704
        Total committed heap usage (bytes)=534249472
    Shuffle Errors
        BAD_ID=0
        CONNECTION=0
        IO_ERROR=0
        WRONG_LENGTH=0
        WRONG_MAP=0
        WRONG_REDUCE=0
    File Input Format Counters
        Bytes Read=2836
    File Output Format Counters
        Bytes Written=1578
    rmr
        reduce calls=3
15/03/12 23:30:53 INFO streaming.StreamJob: Output directory: /tmp/ file5a996d3831d9
    packageJobJar:   [/usr/local/hadoop-2.5.1/tmp/hadoop-unjar3734383471630720759/]   []
/tmp/streamjob38362385101174231620.jar tmpDir=null
    15/03/12 23:31:02 INFO client.RMProxy: Connecting to ResourceManager at/0.0.0.0:8032
    15/03/12 23:31:03 INFO client.RMProxy: Connecting to ResourceManager at/0.0.0.0:8032
    15/03/12 23:31:03 INFO mapred.FileInputFormat: Total input paths to process : 1 15/03/12
23:31:03 INFO mapreduce.JobSubmitter: number of splits:2
    15/03/12   23:31:04   INFO   mapreduce.JobSubmitter:   Submitting   tokens   for   job:
job_1426206027775_0024
    15/03/12 23:31:04 INFO impl.YarnClientImpl: Submitted application
application_1426206027775_0024
    15/03/12 23:31:04 INFO mapreduce.Job: The url to track the job: http://
master:8088/proxy/application_1426206027775_0024/
    15/03/12 23:31:04 INFO mapreduce.Job: Running job: job_1426206027775_0024 15/03/12
23:31:10 INFO mapreduce.Job: Job job_1426206027775_0024 running in uber mode : false
    15/03/12 23:31:10 INFO mapreduce.Job:  map 0% reduce 0%
    15/03/12 23:31:17 INFO mapreduce.Job:  map 50% reduce 0%
    15/03/12 23:31:25 INFO mapreduce.Job:  map 100% reduce 0%
    15/03/12 23:31:31 INFO mapreduce.Job:  map 100% reduce 100%
    15/03/12 23:31:32 INFO mapreduce.Job: Job job_1426206027775_0024 completed successfully
```

```
15/03/12 23:31:32 INFO mapreduce.Job: Counters: 50
        File System Counters
            FILE: Number of bytes read=7294
            FILE: Number of bytes written=321712
            FILE: Number of read operations=0
            FILE: Number of large read operations=0
            FILE: Number of write operations=0
            HDFS: Number of bytes read=3018
            HDFS: Number of bytes written=1578
            HDFS: Number of read operations=13
            HDFS: Number of large read operations=0
            HDFS: Number of write operations=2
        Job Counters
            Launched map tasks=2
            Launched reduce tasks=1
            Data-local map tasks=2
            Total time spent by all maps in occupied slots (ms)=39584
            Total time spent by all reduces in occupied slots (ms)=20296
            Total time spent by all map tasks (ms)=9896
            Total time spent by all reduce tasks (ms)=5074
            Total vcore-seconds taken by all map tasks=9896
            Total vcore-seconds taken by all reduce tasks=5074
            Total megabyte-seconds taken by all map tasks=40534016
            Total megabyte-seconds taken by all reduce tasks=20783104
        Map-Reduce Framework
            Map input records=3
            Map output records=7
            Map output bytes=7260
            Map output materialized bytes=7300
            Input split bytes=182
            Combine input records=0
            Combine output records=0
            Reduce input groups=3
            Reduce shuffle bytes=7300
            Reduce input records=7
            Reduce output records=6
            Spilled Records=14
            Shuffled Maps =2
            Failed Shuffles=0
            Merged Map outputs=2
            GC time elapsed (ms)=232
            CPU time spent (ms)=3110
            Physical memory (bytes) snapshot=678764544
            Virtual memory (bytes) snapshot=3070926848
            Total committed heap usage (bytes)=536870912
        Shuffle Errors
            BAD_ID=0
            CONNECTION=0
            IO_ERROR=0
            WRONG_LENGTH=0
```

```
            WRONG_MAP=0
            WRONG_REDUCE=0
        File Input Format Counters
            Bytes Read=2836
        File Output Format Counters
            Bytes Written=1578
        rmr
            reduce calls=3
15/03/12 23:31:32 INFO streaming.StreamJob: Output directory: /tmp/ file5a9930898e21
    15/03/12 23:31:38 INFO fs.TrashPolicyDefault: Namenode trash configuration: Deletion interval = 0 minutes, Emptier interval = 0 minutes.
    Deleted /tmp/file5a996d3831d9
    packageJobJar:    [/usr/local/hadoop-2.5.1/tmp/hadoop-unjar1311637911573935464/]    []
/tmp/streamjob6945465899804277396.jar tmpDir=null
    15/03/12 23:31:47 INFO client.RMProxy: Connecting to ResourceManager at/0.0.0.0:8032
    15/03/12 23:31:47 INFO client.RMProxy: Connecting to ResourceManager at/0.0.0.0:8032
    15/03/12 23:31:48 INFO mapred.FileInputFormat: Total input paths to process : 1 15/03/12 23:31:48 INFO mapreduce.JobSubmitter: number of splits:2
    15/03/12  23:31:49  INFO  mapreduce.JobSubmitter:  Submitting  tokens  for  job: job_1426206027775_0025
    15/03/12   23:31:49   INFO   impl.YarnClientImpl:   Submitted   application application_1426206027775_0025
    15/03/12  23:31:49  INFO  mapreduce.Job:  The  url  to  track  the  job:  http:// master:8088/proxy/application_1426206027775_0025/
    15/03/12 23:31:49 INFO mapreduce.Job: Running job: job_1426206027775_0025 15/03/12 23:31:55 INFO mapreduce.Job: Job job_1426206027775_0025 running in uber mode : false
    15/03/12 23:31:55 INFO mapreduce.Job:  map 0% reduce 0%
    15/03/12 23:32:02 INFO mapreduce.Job:  map 50% reduce 0%
    15/03/12 23:32:08 INFO mapreduce.Job:  map 100% reduce 0%
    15/03/12 23:32:16 INFO mapreduce.Job:  map 100% reduce 100%
    15/03/12 23:32:16 INFO mapreduce.Job: Job job_1426206027775_0025 completed successfully
    15/03/12 23:32:16 INFO mapreduce.Job: Counters: 50
        File System Counters
            FILE: Number of bytes read=7294
            FILE: Number of bytes written=321682
            FILE: Number of read operations=0
            FILE: Number of large read operations=0
            FILE: Number of write operations=0
            HDFS: Number of bytes read=3018
            HDFS: Number of bytes written=1578
            HDFS: Number of read operations=13
            HDFS: Number of large read operations=0
            HDFS: Number of write operations=2
        Job Counters
            Launched map tasks=2
            Launched reduce tasks=1
            Data-local map tasks=2
            Total time spent by all maps in occupied slots (ms)=37840
            Total time spent by all reduces in occupied slots (ms)=18948
            Total time spent by all map tasks (ms)=9460
```

```
        Total time spent by all reduce tasks (ms)=4737
        Total vcore-seconds taken by all map tasks=9460
        Total vcore-seconds taken by all reduce tasks=4737
        Total megabyte-seconds taken by all map tasks=38748160
        Total megabyte-seconds taken by all reduce tasks=19402752
    Map-Reduce Framework
        Map input records=3
        Map output records=7
        Map output bytes=7260
        Map output materialized bytes=7300
        Input split bytes=182
        Combine input records=0
        Combine output records=0
        Reduce input groups=3
        Reduce shuffle bytes=7300
        Reduce input records=7
        Reduce output records=6
        Spilled Records=14
        Shuffled Maps =2
        Failed Shuffles=0
        Merged Map outputs=2
        GC time elapsed (ms)=165
        CPU time spent (ms)=2750
        Physical memory (bytes) snapshot=712826880
        Virtual memory (bytes) snapshot=3088670720
        Total committed heap usage (bytes)=534249472
    Shuffle Errors
        BAD_ID=0
        CONNECTION=0
        IO_ERROR=0
        WRONG_LENGTH=0
        WRONG_MAP=0
        WRONG_REDUCE=0
    File Input Format Counters
        Bytes Read=2836
    File Output Format Counters
        Bytes Written=1578
    rmr
        reduce calls=3
15/03/12 23:32:16 INFO streaming.StreamJob: Output directory: /tmp/ file5a994566eb31
15/03/12 23:32:21 INFO fs.TrashPolicyDefault: Namenode trash configuration: Deletion interval = 0 minutes, Emptier interval = 0 minutes.
Deleted /tmp/file5a9930898e21
packageJobJar:   [/usr/local/hadoop-2.5.1/tmp/hadoop-unjar5505210954969057280/]   []   /tmp/streamjob5411908344658597.jar tmpDir=null
15/03/12 23:32:30 INFO client.RMProxy: Connecting to ResourceManager at/0.0.0.0:8032
15/03/12 23:32:30 INFO client.RMProxy: Connecting to ResourceManager at/0.0.0.0:8032
15/03/12 23:32:31 INFO mapred.FileInputFormat: Total input paths to process : 1 15/03/12 23:32:32 INFO mapreduce.JobSubmitter: number of splits:2
15/03/12 23:32:32 INFO mapreduce.JobSubmitter: Submitting tokens for job:
```

```
job_1426206027775_0026
    15/03/12    23:32:32    INFO    impl.YarnClientImpl:    Submitted    application
application_1426206027775_0026
    15/03/12  23:32:32  INFO  mapreduce.Job:  The  url  to  track  the  job:  http://
master:8088/proxy/application_1426206027775_0026/
    15/03/12 23:32:32 INFO mapreduce.Job: Running job: job_1426206027775_0026 15/03/12
23:32:39 INFO mapreduce.Job: Job job_1426206027775_0026 running in uber mode : false
    15/03/12 23:32:39 INFO mapreduce.Job:  map 0% reduce 0%
    15/03/12 23:32:46 INFO mapreduce.Job:  map 50% reduce 0%
    15/03/12 23:32:51 INFO mapreduce.Job:  map 100% reduce 0%
    15/03/12 23:32:58 INFO mapreduce.Job:  map 100% reduce 100%
    15/03/12 23:32:58 INFO mapreduce.Job: Job job_1426206027775_0026 completed successfully
    15/03/12 23:32:58 INFO mapreduce.Job: Counters: 50
        File System Counters
                FILE: Number of bytes read=7294
                FILE: Number of bytes written=321715
                FILE: Number of read operations=0
                FILE: Number of large read operations=0
                FILE: Number of write operations=0
                HDFS: Number of bytes read=3018
                HDFS: Number of bytes written=1578
                HDFS: Number of read operations=13
                HDFS: Number of large read operations=0
                HDFS: Number of write operations=2
        Job Counters
                Launched map tasks=2
                Launched reduce tasks=1
                Data-local map tasks=2
                Total time spent by all maps in occupied slots (ms)=36796
                Total time spent by all reduces in occupied slots (ms)=21436
                Total time spent by all map tasks (ms)=9199
                Total time spent by all reduce tasks (ms)=5359
                Total vcore-seconds taken by all map tasks=9199
                Total vcore-seconds taken by all reduce tasks=5359
                Total megabyte-seconds taken by all map tasks=37679104
                Total megabyte-seconds taken by all reduce tasks=21950464
        Map-Reduce Framework
                Map input records=3
                Map output records=7
                Map output bytes=7260
                Map output materialized bytes=7300
                Input split bytes=182
                Combine input records=0
                Combine output records=0
                Reduce input groups=3
                Reduce shuffle bytes=7300
                Reduce input records=7
                Reduce output records=6
                Spilled Records=14
                Shuffled Maps =2
```

```
            Failed Shuffles=0
            Merged Map outputs=2
            GC time elapsed (ms)=123
            CPU time spent (ms)=3080
            Physical memory (bytes) snapshot=701104128
            Virtual memory (bytes) snapshot=3083026432
            Total committed heap usage (bytes)=536870912
        Shuffle Errors
            BAD_ID=0
            CONNECTION=0
            IO_ERROR=0
            WRONG_LENGTH=0
            WRONG_MAP=0
            WRONG_REDUCE=0
        File Input Format Counters
            Bytes Read=2836
        File Output Format Counters
            Bytes Written=1578
        rmr
            reduce calls=3
    15/03/12 23:32:58 INFO streaming.StreamJob: Output directory: /tmp/ file5a997c96e3da
    15/03/12 23:33:03 INFO fs.TrashPolicyDefault: Namenode trash configuration: Deletion
interval = 0 minutes, Emptier interval = 0 minutes.
    Deleted /tmp/file5a994566eb31
    packageJobJar:   [/usr/local/hadoop-2.5.1/tmp/hadoop-unjar3406204528169182264/]   []
/tmp/streamjob6381189447976749478.jar tmpDir=null
    15/03/12 23:33:12 INFO client.RMProxy: Connecting to ResourceManager at /0.0.0.0:8032
    15/03/12 23:33:13 INFO client.RMProxy: Connecting to ResourceManager at /0.0.0.0:8032
    15/03/12 23:33:13 INFO mapred.FileInputFormat: Total input paths to process : 1 15/03/12
23:33:13 INFO mapreduce.JobSubmitter: number of splits:2
    15/03/12  23:33:14  INFO  mapreduce.JobSubmitter:  Submitting  tokens  for  job:
job_1426206027775_0027
    15/03/12    23:33:14    INFO    impl.YarnClientImpl:    Submitted    application
application_1426206027775_0027
    15/03/12  23:33:14  INFO  mapreduce.Job:  The  url  to  track  the  job:  http://
master:8088/proxy/application_1426206027775_0027/
    15/03/12 23:33:14 INFO mapreduce.Job: Running job: job_1426206027775_0027 15/03/12
23:33:21 INFO mapreduce.Job: Job job_1426206027775_0027 running in uber mode : false
    15/03/12 23:33:21 INFO mapreduce.Job:  map 0% reduce 0%
    15/03/12 23:33:28 INFO mapreduce.Job:  map 50% reduce 0%
    15/03/12 23:33:34 INFO mapreduce.Job:  map 100% reduce 0%
    15/03/12 23:33:40 INFO mapreduce.Job:  map 100% reduce 100%
    15/03/12 23:33:40 INFO mapreduce.Job: Job job_1426206027775_0027 completed successfully
    15/03/12 23:33:40 INFO mapreduce.Job: Counters: 50
        File System Counters
            FILE: Number of bytes read=7294
            FILE: Number of bytes written=321721
            FILE: Number of read operations=0
            FILE: Number of large read operations=0
            FILE: Number of write operations=0
```

```
        HDFS: Number of bytes read=3018
        HDFS: Number of bytes written=1578
        HDFS: Number of read operations=13
        HDFS: Number of large read operations=0
        HDFS: Number of write operations=2
    Job Counters
        Launched map tasks=2
        Launched reduce tasks=1
        Data-local map tasks=2
        Total time spent by all maps in occupied slots (ms)=37844
        Total time spent by all reduces in occupied slots (ms)=19440
        Total time spent by all map tasks (ms)=9461
        Total time spent by all reduce tasks (ms)=4860
        Total vcore-seconds taken by all map tasks=9461
        Total vcore-seconds taken by all reduce tasks=4860
        Total megabyte-seconds taken by all map tasks=38752256
        Total megabyte-seconds taken by all reduce tasks=19906560
    Map-Reduce Framework
        Map input records=3
        Map output records=7
        Map output bytes=7260
        Map output materialized bytes=7300
        Input split bytes=182
        Combine input records=0
        Combine output records=0
        Reduce input groups=3
        Reduce shuffle bytes=7300
        Reduce input records=7
        Reduce output records=6
        Spilled Records=14
        Shuffled Maps =2
        Failed Shuffles=0
        Merged Map outputs=2
        GC time elapsed (ms)=195
        CPU time spent (ms)=2970
        Physical memory (bytes) snapshot=708616192
        Virtual memory (bytes) snapshot=3084320768
        Total committed heap usage (bytes)=531628032
    Shuffle Errors
        BAD_ID=0
        CONNECTION=0
        IO_ERROR=0
        WRONG_LENGTH=0
        WRONG_MAP=0
        WRONG_REDUCE=0
    File Input Format Counters
        Bytes Read=2836
    File Output Format Counters
        Bytes Written=1578
rmr
```

```
            reduce calls=3
    15/03/12 23:33:40 INFO streaming.StreamJob: Output directory: /tmp/ file5a99c3f154e
    15/03/12 23:33:45 INFO fs.TrashPolicyDefault: Namenode trash configuration: Deletion
interval = 0 minutes, Emptier interval = 0 minutes.
    Deleted /tmp/file5a997c96e3da
```

输出结果:

```
> output
     Sepal.Length Sepal.Width Petal.Length Petal.Width
[1,]    6.628333    3.001667      5.4150      1.930
[2,]    5.006000    3.428000      1.4620      0.246
[3,]    5.712500    2.677500      4.1425      1.295
```

程序范例 13-5

先设置 R+Hadoop 的环境目录:

```
> Sys.setenv(
+    HIVE_HOME= '/usr/local/hive',
+    HADOOP_HOME='/usr/local/hadoop-2.5.1',
+    HADOOP_CMD='/usr/local/hadoop-2.5.1/bin/hadoop',
+    HADOOP_CONF_DIR='/usr/local/hadoop-2.5.1/etc/hadoop',
+    HADOOP_STREAMING      ='/usr/local/hadoop-2.5.1/share/hadoop/tools/lib/hadoop-streaming-2.5.1.jar')
```

使用 rhdfs、rmr2、rhbase、RHive 程序包并启用 hdfs:

```
> library('rhdfs')
Loading required package: rJava

HADOOP_CMD=/usr/local/hadoop-2.5.1/bin/hadoop
Be sure to run hdfs.init()

> library('rmr2')
> library('rhbase')
> library('RHive')
```

启用 rhdfs 及 rhbase:

```
> hdfs.init()
> hb.init()
```

启用 rhive:

```
>            rhive.init(hiveHome="/usr/local/hive",hiveLib="/usr/local/hive/lib",hadoopHome="/usr/local/hadoop-2.5.1")
> rhive.connect('master', hiveServer2=TRUE)
15/03/15 08:37:55 INFO Configuration.deprecation: fs.default.name is deprecated. Instead, use fs.defaultFS
    15/03/15 08:38:02 INFO jdbc.Utils: Supplied authorities: master:10000 15/03/15 08:38:02 INFO jdbc.Utils: Resolved authority: master:10000
```

```
15/03/15 08:38:02 INFO jdbc.HiveConnection: Will try to open client transport with JDBC
Uri: jdbc:hive2://master:10000/default
```

查询数据库名称（使用默认值）：

```
> rhive.query('show databases')
  database_name
1       default
```

查询数据表名称（尚未创建）：

```
> rhive.list.tables()
[1] tab_name
<0 rows> (or 0-length row.names)
```

从 GDP.csv 读取数据并转换 myData 为数据框对象（用户可下载 WinSCP 免费软件并把 GDP.csv 上传至 /home/hadoop）：

```
> myData=read.csv("GDP.csv",header=FALSE)
> myData=as.data.frame(myData)
> myData
    V1 V2                 V3       V4
1  USA  1      United States 16244600
2  CHN  2              China  8227103
3  JPN  3              Japan  5961066
4  DEU  4            Germany  3425928
5  FRA  5             France  2611200
6  GBR  6     United Kingdom  2475782
7  BRA  7             Brazil  2252664
8  RUS  8 Russian Federation  2014775
9  ITA  9              Italy  2013375
10 IND 10              India     1858
```

可使用 rhive.write.table() 函数创建数据表（test_table）并将 myData 写入：

```
> rhive.write.table(myData,'test_table')
[1] "test_table"
```

查询 test_table 数据表的数据结构：

```
> rhive.desc.table('test_table')
  col_name data_type comment
1       v1    string
2       v2       int
3       v3    string
4       v4       int
```

查询 test_table 数据表的数据：

```
> df1=rhive.query('SELECT * from test_table')
> df1
```

```
   test_table.v1 test_table.v2    test_table.v3 test_table.v4
1              USA              1      United States       16244600
2              CHN              2              China        8227103
3              JPN              3              Japan        5961066
4              DEU              4            Germany        3425928
5              FRA              5             France        2611200
6              GBR              6     United Kingdom        2475782
7              BRA              7             Brazil        2252664
8              RUS              8 Russian Federation        2014775
9              ITA              9              Italy        2013375
10             IND             10              India           1858
```

使用 wordcloud 程序包：

```
> library(wordcloud)
Loading required package: RColorBrewer
```

绘出文字云（见图 13-4）：

```
> m1=as.matrix(df1[,4])
> words1=as.matrix(df1[,3])
> wordcloud(words1, m1)
```

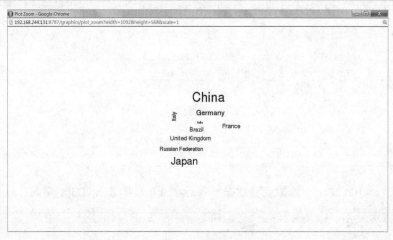

图 13-4　test_table 的文字云

删除 test_table 数据表：

```
> rhive.drop.table('test_table')
```

程序范例 13-6

本范例先使用 Sqoop 导入 MS SQL 数据库及其数据表，再使用 RHive 读取数据表的数据，系统架构如图 13-5 所示。

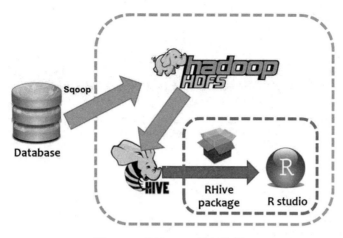

图 13-5　范例 13-6 的架构图

读者可先将本书的 GDPSQL 数据库附加到 MS SQL Server 中（注意要将防火墙 port1433 打开并在 SQL Server 配置管理器中启用 TCP/IP），接着修改并执行 R+Hadoop 平台的 /home/Hadoop/sq.sh。

读者可先使用 Putty 软件以 root 账号登录 R+Hadoop 平台并切换为 hadoop 账号的权限（Putty 软件免费下载网址：http://www.chiark.greenend.org.uk/~sgtatham/putty/download.html）：

```
login as:
root root@192.168.244.131's password:
Last login: Tue Mar 24 23:14:38 2015 from 192.168.244.131

[root@master ~]# su - hadoop
Last login: Thu Mar 26 21:20:46 EDT 2015
```

确认有 sq.sh 执行文件：

```
[hadoop@master ~]$ ls
All_R       Music      Videos    derby.log      nohup.out    scala-2.9.3
Desktop     ictures    hadoop-2.5.1.tar.gz  old_files  scala-2.9.3.tgz
Documents   Public     apache-hive-0.14.0-bin  hadoop_demo.R
Downloads   R  apache-hive-0.14.0-bin.tar.gz  hist.java    rhive_init_    in_rstudio
spark-1.1.1-bin-hadoop2.4.tgz
GDP.java  Templates   caacce.javalibrary_1.csv  rhive_start.init sq.sh
```

修改 sq.sh 相关参数，其中 TB_NAME 表示数据表名称，SQL_SRV 表示 MS SQL Server 的网址，P 表示 sa 的密码，database=GDPSQL 表示数据库名称，sqoop import --connect "jdbc:sqlserver://$SQL_SRV:$PORT;database=GDPS QL;username=$U;password=$P" --hive-import -m 1 --table $i --warehouse- dir $DEST_DIR --hive-overwrite 就是导入 GDPSQL 数据库及其 GDP 数据表：

```
[hadoop@master ~]$ vi sq.sh

#!/bin/bash
TB_NAME='GDP'
```

```
SQL_SRV=192.168.28.1
PORT=1433
U=sa
P='123456'
DEST_DIR=/user/hive/warehouse/mitopac HI_HOME=/usr/local/hive

HIVE_PID=`ps -ef | grep HiveServer2 | awk '{print $2}' | head -1` kill -9 $HIVE_PID

sleep 2

for i in $TB_NAME
do
sqoop     import     --connect    "jdbc:sqlserver://$SQL_SRV:$PORT;database=GDPSQL;username=$U;password=$P" --hive-import -m 1 --table $i --warehouse-dir $DEST_DIR
--hive-overwrite done

sleep 2
nohup $HI_HOME/bin/hive --service hiveserver2 &
```

执行 sq.sh（注意需按 Enter 键结束导入程序）：

```
[hadoop@master ~]$ ./sq.sh
    Warning: /usr/local/sqoop/../hcatalog does not exist! HCatalog jobs will fail. Please set
$HCAT_HOME to the root of your HCatalog installation.
    Warning: /usr/local/sqoop/../accumulo does not exist! Accumulo imports will fail.
    Please set $ACCUMULO_HOME to the root of your Accumulo installation. Warning:
/usr/local/sqoop/../zookeeper does not exist! Accumulo imports will fail.
    Please set $ZOOKEEPER_HOME to the root of your Zookeeper installation. 15/03/26 22:04:47 INFO
sqoop.Sqoop: Running Sqoop version: 1.4.5
    15/03/26 22:04:47 INFO tool.BaseSqoopTool: Using Hive-specific delimiters for output. You can
override
    15/03/26 22:04:47 INFO tool.BaseSqoopTool: delimiters with --fields- terminated-by, etc.
    15/03/26 22:04:47 WARN tool.BaseSqoopTool: It seems that you're doing hive import directly
into default
    15/03/26 22:04:47 WARN tool.BaseSqoopTool: hive warehouse directory which is not supported.
Sqoop is
    15/03/26 22:04:47 WARN tool.BaseSqoopTool: firstly importing data into separate directory and
then
    15/03/26 22:04:47 WARN tool.BaseSqoopTool: inserting data into hive. Please consider removing
    15/03/26  22:04:47  WARN  tool.BaseSqoopTool:  --target-dir  or  --warehouse-dir  into
/user/hive/warehouse in
    15/03/26 22:04:47 WARN tool.BaseSqoopTool: case that you will detect any issues.
    15/03/26 22:04:48 INFO manager.SqlManager: Using default fetchSize of 1000 15/03/26 22:04:48
INFO tool.CodeGenTool: Beginning code generation
    15/03/26 22:04:49 INFO manager.SqlManager: Executing SQL statement: SELECT t.* FROM [GDP] AS
t WHERE 1=0
    15/03/26 22:04:50 INFO orm.CompilationManager: HADOOP_MAPRED_HOME is /usr/ local/hadoop-2.5.1
    Note: /tmp/sqoop-hadoop/compile/30962d4742ac9200621ca139f3b22f91/GDP.java uses or overrides
a deprecated API.
    Note: Recompile with -Xlint:deprecation for details.
    15/03/26   22:04:56   INFO   orm.CompilationManager:   Writing   jar   file:   /tmp/sqoop-
```

```
hadoop/compile/30962d4742ac9200621ca139f3b22f91/GDP.jar
    15/03/26 22:04:57 INFO mapreduce.ImportJobBase: Beginning import of GDP 15/03/26 22:04:57 INFO
Configuration.deprecation: mapred.jar is deprecated. Instead, use mapreduce.job.jar
    15/03/26 22:04:59 INFO Configuration.deprecation: mapred.map.tasks is deprecated. Instead,
use mapreduce.job.maps
    15/03/26 22:04:59 INFO client.RMProxy: Connecting to ResourceManager at
    /0.0.0.0:8032
    15/03/26 22:05:04 INFO db.DBInputFormat: Using read commited transaction isolation
    15/03/26 22:05:05 INFO mapreduce.JobSubmitter: number of splits:1 15/03/26 22:05:05 INFO
mapreduce.JobSubmitter: Submitting tokens for job: job_1427419310304_0001
    15/03/26    22:05:07    INFO    impl.YarnClientImpl:    Submitted    application
application_1427419310304_0001
    15/03/26 22:05:07 INFO mapreduce.Job: The url to track the job: http://
master:8088/proxy/application_1427419310304_0001/
    15/03/26 22:05:07 INFO mapreduce.Job: Running job: job_1427419310304_0001 15/03/26 22:05:27
INFO mapreduce.Job: Job job_1427419310304_0001 running in uber mode : false
    15/03/26 22:05:27 INFO mapreduce.Job:  map 0% reduce 0%
    15/03/26 22:05:40 INFO mapreduce.Job:  map 100% reduce 0%
    15/03/26 22:05:41 INFO mapreduce.Job: Job job_1427419310304_0001 completed successfully
    15/03/26 22:05:42 INFO mapreduce.Job: Counters: 30
            File System Counters
                FILE: Number of bytes read=0
                FILE: Number of bytes written=105805
                FILE: Number of read operations=0
                FILE: Number of large read operations=0
                FILE: Number of write operations=0
                HDFS: Number of bytes read=87
                HDFS: Number of bytes written=233
                HDFS: Number of read operations=4
                HDFS: Number of large read operations=0
                HDFS: Number of write operations=2
            Job Counters
                Launched map tasks=1
                Other local map tasks=1
                Total time spent by all maps in occupied slots (ms)=11617
                Total time spent by all reduces in occupied slots (ms)=0
                Total time spent by all map tasks (ms)=11617
                Total vcore-seconds taken by all map tasks=11617
                Total megabyte-seconds taken by all map tasks=11895808
            Map-Reduce Framework
                Map input records=10
                Map output records=10
                Input split bytes=87
                Spilled Records=0
                Failed Shuffles=0
                Merged Map outputs=0
                GC time elapsed (ms)=292
                CPU time spent (ms)=3520
                Physical memory (bytes) snapshot=182464512
                Virtual memory (bytes) snapshot=810999808
```

```
            Total committed heap usage (bytes)=106954752
        File Input Format Counters
            Bytes Read=0
        File Output Format Counters
            Bytes Written=233
    15/03/26 22:05:42 INFO mapreduce.ImportJobBase: Transferred 233 bytes in 43.1651 seconds
(5.3979 bytes/sec)
    15/03/26 22:05:42 INFO mapreduce.ImportJobBase: Retrieved 10 records. 15/03/26 22:05:42 INFO
manager.SqlManager: Executing SQL statement: SELECT t.* FROM [GDP] AS t WHERE 1=0
    15/03/26 22:05:42 INFO hive.HiveImport: Loading uploaded data into Hive 15/03/26 22:05:54
INFO hive.HiveImport:
    15/03/26 22:05:54 INFO hive.HiveImport: Logging initialized using configuration in
file:/usr/local/apache-hive-0.14.0-bin/conf/hive-log4j.properties
    15/03/26 22:05:55 INFO hive.HiveImport: SLF4J: Class path contains multiple SLF4J bindings.
    15/03/26 22:05:55 INFO hive.HiveImport: SLF4J: Found binding in [jar:file:/
usr/local/hadoop-2.5.1/share/hadoop/common/lib/slf4j-log4j12-1.7.5.jar!/org/
slf4j/impl/StaticLoggerBinder.class]
    15/03/26 22:05:55 INFO hive.HiveImport: SLF4J: Found binding in [jar:file:/
usr/local/apache-hive-0.14.0-bin/lib/hive-jdbc-0.14.0-standalone.jar!/org/
slf4j/impl/StaticLoggerBinder.class]
    15/03/26 22:05:55 INFO hive.HiveImport: SLF4J: See http://www.slf4j.org/codes.
html#multiple_bindings for an explanation.
    15/03/26 22:05:55 INFO hive.HiveImport: SLF4J: Actual binding is of type [org.
slf4j.impl.Log4jLoggerFactory]
    15/03/26 22:06:17 INFO hive.HiveImport: OK
    15/03/26 22:06:17 INFO hive.HiveImport: Time taken: 3.85 seconds 15/03/26 22:06:18 INFO
hive.HiveImport: Loading data to table default.gdp
    15/03/26 22:06:19 INFO hive.HiveImport: Table default.gdp stats: [numFiles=1, numRows=0,
totalSize=233, rawDataSize=0]
    15/03/26 22:06:19 INFO hive.HiveImport: OK
    15/03/26 22:06:19 INFO hive.HiveImport: Time taken: 2.005 seconds 15/03/26 22:06:20 INFO
hive.HiveImport: Hive import complete. [hadoop@master ~]$ nohup: appending output to
'nohup.out'
```

导入 GDPSQL 数据库后，用户可使用 Google Chrome 连接到 RStudio，在 RStudio 中设置 R+Hadoop 的环境目录：

```
> ## Required path for R+Hadoop
> Sys.setenv(
+    HIVE_NAME= '/usr/local/hive',
+    HADOOP_HOME='/usr/local/hadoop-2.5.1',
+    HADOOP_CMD='/usr/local/hadoop-2.5.1/bin/hadoop',
+    HADOOP_CONF_DIR='/usr/local/hadoop-2.5.1/etc/hadoop',
+    HADOOP_STREAMING          ='/usr/local/hadoop-2.5.1/share/hadoop/tools/lib/hadoop-streaming-2.5.1.jar')
```

使用 rhdfs、rmr2、rhbase、RHive 程序包并启用 hdfs：

```
> library('rhdfs')
Loading required package: rJava
```

```
HADOOP_CMD=/usr/local/hadoop-2.5.1/bin/hadoop

Be sure to run hdfs.init()
> library('rmr2')
> library('rhbase')
> library('RHive')

> hdfs.init()
```

启用 rhive:

```
> rhive.init(hiveHome="/usr/local/hive",hiveLib="/usr/local/hive/lib",hadoopHome="/usr/local/hadoop-2.5.1")
> rhive.connect('master', hiveServer2=TRUE)
15/03/26 22:30:06 INFO Configuration.deprecation: fs.default.name is deprecated. Instead, use fs.defaultFS
15/03/26 22:30:09 INFO jdbc.Utils: Supplied authorities: master:10000
15/03/26 22:30:09 INFO jdbc.Utils: Resolved authority: master:10000
15/03/26 22:30:09 INFO jdbc.HiveConnection: Will try to open client transport with JDBC Uri: jdbc:hive2://master:10000/default
```

查询数据库名称（使用默认值）：

```
> rhive.query('show databases')
 database_name
1       default
```

查询数据表名称：

```
> rhive.list.tables()
 tab_name
1      gdp
```

查询 gdp 数据表的数据：

```
> df1=rhive.query('SELECT * from gdp')
> df1
   gdp.v1 gdp.v2           gdp.v3    gdp.v4
1     USA      1    United States  16244600
2     CHN      2            China   8227103
3     JPN      3            Japan   5961066
4     DEU      4          Germany   3425928
5     FRA      5           France   2611200
6     GBR      6   United Kingdom   2475782
7     BRA      7           Brazil   2252664
8     RUS      8 Russian Federation 2014775
9     ITA      9            Italy   2013375
10    IND     10            India      1858
```

删除 gdp 数据表：

```
> rhive.drop.table('gdp')
```

第 14 章 SparkR 大数据分析

本章重点内容：

- dplyr 数据处理程序包
- SparkR 数据处理
- SparkR 与 SQL Server
- SparkR 与 Cassandra
- Spark Standalone 模式
- SparkR 数据分析

Apache Spark 是一个开放源码集群（Cluster）运算框架，由加州大学伯克利分校 AMPLab 所开发，是一个用于大量数据处理的快速且通用的引擎。Apache Spark 的特色为：指令周期快（Speed）、易于使用（Ease of Use）、具有通用性（Generality）且兼容性高（Runs Everywhere）。Apache Spark 的特色及说明如表 14-1 所示。

表 14-1 Apache Spark 特色及说明

特色	说明
指令周期快	Spark 是基于内存的开放源码集群运算系统，比 Hadoop MapReduce 快 100 倍。即便是在硬盘上执行程序，Spark 的速度也能快 10 倍
易于使用	支持 Scala、Python 及 Java 程序设计语言，并在 1.4 版本后可支持 R 语言，使得开发者可以根据应用的环境决定所需语言来开发
具有通用性	可结合 SQL、Streaming 及复杂分析（Analytics）功能
兼容性高	Spark 可在 Hadoop、Mesos、Standalone 或云（Cloud）上运行，可存取 HDFS、Cassandra、Hbase 及 S3 等数据

Apache Spark 包含 Spark SQL、Spark Streaming、MLlib 及 GraphX 模块，其架构图如图 14-1 所示。

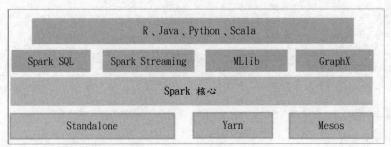

图 14-1 Apache Spark 架构图

Spark SQL 主要用于结构化数据的 Apache Spark 模块,具有集成性(Integrated),可无缝结合 SQL Queries 与 Spark 程序;统一的数据存取(Uniform Data Access)能以相同方式连接数据;与 Hive 兼容(Compatibility),不需修改 Hive Queries 程序;具有标准化连通性(Standard Connectivity),可支持 ODBC、JDBC。Spark Streaming 是 Spark 核心 API 的一个扩展,它对实时数据串流的处理具有可扩展性、高吞吐量、可容错性等特点,让创建串流应用程序更容易,其易于使用,可创建高级操作应用程序(High-Level Operators);具有容错性(Fault Tolerance)及状态性(Stateful);可与 Spark 集成(Integration),具有批次(Batch)和交互(Interactive)查询功能。MLlib 是 Apache Spark 上可扩展的机器学习函数库,可使用 Java、Scala、Python 及 R 语言;具有高性能(Performance),比 MapReduce 快 100 倍;容易部署(Deploy),可在 Hadoop 上运行。GraphX 是 Apache Spark 用于图形和并行图形计算的 API,具有弹性(Flexibility),速度快且提供多种算法。

用户可在 http://spark.apache.org/downloads.html 下载最新版的 Spark。Spark 执行模式可分为 Local、Standalone、Yarn 和 Mesos,其安装方式可参考附录 F。Local 模式就是单机版,适合单机测试及在自己的节点上学习 Spark。Standalone 是 Spark 的 Client-Server 应用程序,设置好 Slaves 文件后,可发布到各个节点上创建集群环境。Mesos 与 Yarn 主要是把 Standalone 模式在资源管理框架上执行,让资源管理框架弹性地分配运算资源。一般来说,执行简单测试时可先使用 Local 模式运行。不需要资源规划分和分配优先等级时,可以使用 Standalone 模式来部署。希望可以限制固定运行的最大资源,或者需要按优先等级运行时,可以选择 Yarn 或 Mesos。

Spark 的核心是弹性分布式数据集(Resilient Distributed Dataset,RDD),是由 AMPLab 实验室提出的一种分布式内存概念。RDD 能与其他系统兼容,可以导入外部存储系统的数据集,例如 HDFS、HBase 或其他 Hadoop 数据来源。Spark RDD 的架构如图 14-2 所示,SparkContext 是 Spark 的主程序(Driver Program),Cluster 集群中包含多台 Worker Node,Worker Node 是实际执行程序的计算机,执行时会将数据存储在 RDD 内存中。Cluster Manager 将 RDD 自动分为多个 Partition,分别在不同的 Worker Node 计算机中执行,执行结果再回传到 SparkContext。

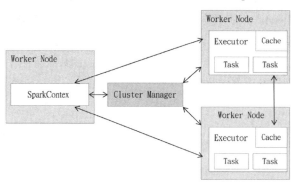

图 14-2　Spark RDD 架构图

RDD 的主要特性为 Immutable 和 Lineage 机制。Immutable 具有不可改变的特性,可使 RDD 具备容错的特性,即使特定节点损毁,存储于其上的 RDD 也能重新计算得出,因而避免了特定节点损毁就无法运行的问题;Lineage 记录了每个 RDD 与其父代 RDD 之间的关联。Lineage 的机制如图 14-3 所示。

图 14-3　Lineage 的机制

输入数据，执行"转换 1"运算产生 RDD1，此时不会实际执行，只记录操作命令，RDD1 执行"转换 2"运算而产生 RDD2，此时也不会实际执行，只记录操作命令，RDD2 执行"动作 1"运算，此时才会实际执行来产生输出数据。RDD 运算类型共分 3 种，如表 14-2 所示。

表 14-2　RDD 运算的三种类型

RDD 运算类型	说明
转换（Transformation）	RDD 执行转换运算后，会产生另一个 RDD，而在还未执行动作时，不会实际执行转换运算
动作（Action）	RDD 执行动作运算后，不会产生另一个 RDD，而是产生数值、数组或写入文件系统。执行时会连同之前的转换运算一起执行
持久化（Persistence）	对于重复使用的 RDD，可将其持久化在内存中，以便后续使用时提升性能

然而，RDD 有不易使用和无结构描述（Schema）的缺点，为了解决此问题，有了 Spark DataFrame 和 Spark SQL，用户可以先使用 createDataFrame()或 read.df()函数创建 Spark DataFrame 对象，再使用类似 dplyr 程序包的数据处理功能，也可以通过 createOrReplaceTempView()转换为 Spark SQL TABLE 来使用 Spark SQL 语法。

14.1　dplyr 数据处理程序包

使用 SparkR 之前，先介绍相关的 dplyr 程序包，让用户先熟悉数据处理的功能。dplyr 提供处理数据的函数主要有 select()、filter()、arrange()、mutate()、group_by()、n() 和 summarise()。select()函数用于选择特定列（Column）的数据，filter()函数按照自定义的逻辑条件来筛选出符合条件的数据，arrange()函数用于选择特定列的数据来进行排序，mutate() 函数对现有的列数据进行运算并产生新的列数据，group_by()函数用于选择列数据进行分组，n() 函数是分组中数据的项数，summarise()函数是对数据进行统计运算并返回结果。

程序范例 14-1

首先使用 dplyr 程序包和 iris 数据集：

```
> library(dplyr)
> data(iris)
```

使用 select()选择 iris 中 Petal.Length 和 Petal.Width 的列数据并赋值给 sub_iris 对象：

```
> sub_iris <- select(iris, Petal.Length, Petal.Width)
```

显示 sub_iris 对象前 6 项数据：

```
> head(sub_iris)
```

```
  Petal.Length Petal.Width
1     1.4         0.2
2     1.4         0.2
3     1.3         0.2
4     1.5         0.2
5     1.4         0.2
6     1.7         0.4
```

使用 filter()函数筛选出符合 Petal.Length < 1 或 Petal.Width <1 条件的前 6 项数据:

```
> head(filter(iris, Petal.Length < 1 | Petal.Width <1))
  Sepal.Length Sepal.Width Petal.Length Petal.Width Species
1     5.1         3.5         1.4         0.2     setosa
2     4.9         3.0         1.4         0.2     setosa
3     4.7         3.2         1.3         0.2     setosa
4     4.6         3.1         1.5         0.2     setosa
5     5.0         3.6         1.4         0.2     setosa
6     5.4         3.9         1.7         0.4     setosa
```

使用 filter()函数筛选出符合 Petal.Length 为 1.4 而且 Petal.Width 为 0.2 条件的前 6 项数据:

```
> head(filter(iris, Petal.Length == 1.4 & Petal.Width == 0.2))
  Sepal.Length Sepal.Width Petal.Length Petal.Width Species
1     5.1         3.5         1.4         0.2     setosa
2     4.9         3.0         1.4         0.2     setosa
3     5.0         3.6         1.4         0.2     setosa
4     4.4         2.9         1.4         0.2     setosa
5     5.2         3.4         1.4         0.2     setosa
6     5.5         4.2         1.4         0.2     setosa
```

使用 arrange()函数选择 Petal.Length 来进行降序排序并显示前 6 项数据:

```
> head(arrange(iris, desc(Petal.Length)))
  Sepal.Length Sepal.Width Petal.Length Petal.Width  Species
1     7.7         2.6         6.9         2.3     virginica
2     7.7         3.8         6.7         2.2     virginica
3     7.7         2.8         6.7         2.0     virginica
4     7.6         3.0         6.6         2.1     virginica
5     7.9         3.8         6.4         2.0     virginica
6     7.3         2.9         6.3         1.8     virginica
```

使用 arrange() 函数选择 Petal.Length 进行降序排序、Petal.Width 进行升序排序并显示前 6 项数据(以 Petal.Length 先降序排序,相同时再以 Petal.Width 升序排序):

```
> head(arrange(iris, desc(Petal.Length), Petal.Width))
  Sepal.Length Sepal.Width Petal.Length Petal.Width  Species
1     7.7         2.6         6.9         2.3     virginica
2     7.7         2.8         6.7         2.0     virginica
3     7.7         3.8         6.7         2.2     virginica
```

4	7.6	3.0	6.6	2.1	virginica
5	7.9	3.8	6.4	2.0	virginica
6	7.3	2.9	6.3	1.8	virginica

使用 arrange()函数选择 Petal.Width 进行升序排序、Petal.Length 进行降序排序并显示前 6 项数据（以 Petal.Width 先升序排序，相同时再以 Petal.Length 降序排序）：

```
> head(arrange(iris, Petal.Width, desc(Petal.Length)))
  Sepal.Length Sepal.Width Petal.Length Petal.Width Species
1          4.9         3.1          1.5         0.1  setosa
2          5.2         4.1          1.5         0.1  setosa
3          4.8         3.0          1.4         0.1  setosa
4          4.9         3.6          1.4         0.1  setosa
5          4.3         3.0          1.1         0.1  setosa
6          4.8         3.4          1.9         0.2  setosa
```

使用 mutate()函数产生新的 Petal.Length.new 列数据并显示前 6 项数据：

```
> head(mutate(iris, Petal.Length.new = Petal.Length/ 10))
  Sepal.Length Sepal.Width Petal.Length Petal.Width Species Petal.Length.new
1          5.1         3.5          1.4         0.2  setosa             0.14
2          4.9         3.0          1.4         0.2  setosa             0.14
3          4.7         3.2          1.3         0.2  setosa             0.13
4          4.6         3.1          1.5         0.2  setosa             0.15
5          5.0         3.6          1.4         0.2  setosa             0.14
6          5.4         3.9          1.7         0.4  setosa             0.17
```

使用 group_by()函数并依照 Petal.Width 行数据来做群组化并显示前 6 项数据：

```
> by_Petal.Width <- group_by(iris, Petal.Width)
> head(by_Petal.Width)
Source: local data frame [6 x 5] Groups: Petal.Width [2]

  Sepal.Length Sepal.Width Petal.Length Petal.Width Species
         <dbl>       <dbl>        <dbl>       <dbl>  <fctr>
1          5.1         3.5          1.4         0.2  setosa
2          4.9         3.0          1.4         0.2  setosa
3          4.7         3.2          1.3         0.2  setosa
4          4.6         3.1          1.5         0.2  setosa
5          5.0         3.6          1.4         0.2  setosa
6          5.4         3.9          1.7         0.4  setosa
```

使用 summarise()和 n()函数来计算分组的数量（项数）及各分组的平均值：

```
>   summarise(by_Petal.Width, n=n(), mean(Petal.Width))
# A tibble: 22×3
  Petal.Width     n `mean(Petal.Width)`
        <dbl> <int>               <dbl>
1         0.1     5                 0.1
2         0.2    29                 0.2
```

```
3        0.3         7         0.3
4        0.4         7         0.4
5        0.5         1         0.5
6        0.6         1         0.6
7        1.0         7         1.0
8        1.1         3         1.1
9        1.2         5         1.2
10       1.3        13         1.3
# ... with  12 more rows
```

注意，以上两个函数会将结果转换成 tbl/tibble 格式，用户可使用 as.data.frame() 函数转换为数据框。

```
> df_by_Petal.Width <- as.data.frame(by_Petal.Width)
> str(df_by_Petal.Width)
'data.frame':150 obs. of  5 variables:
$ Sepal.Length: num  5.1 4.9 4.7 4.6 5 5.4 4.6 5 4.4 4.9 ...
$ Sepal.Width : num  3.5 3 3.2 3.1 3.6 3.9 3.4 3.4 2.9 3.1 ...
$ Petal.Length: num  1.4 1.4 1.3 1.5 1.4 1.7 1.4 1.5 1.4 1.5 ...
$ Petal.Width : num  0.2 0.2 0.2 0.2 0.2 0.4 0.3 0.2 0.2 0.1 ...
$ Species : Factor w/ 3 levels "setosa","versicolor",..: 1 1 1 1 1 1 1 1 1 1 ...
```

14.2　SparkR 数据处理

读者可参考附录 F 安装 SparkR 并使用浏览器连接到用户设置网址的 RStudio，例如 192.168.244.150:8787。

程序范例 14-2

首先设置 SPARK_HOME 路径和使用 SparkR 程序包：

```
> if (nchar(Sys.getenv("SPARK_HOME")) < 1) {
+    Sys.setenv(SPARK_HOME = "/usr/local/spark")
+ }
> library(SparkR, lib.loc = c(file.path(Sys.getenv("SPARK_HOME"), "R", "lib")))
```

启用 sparkR.session 并设置 appName 名称：

```
> sparkR.session(appName = "SparkR-example")
Java ref type org.apache.spark.sql.SparkSession id 1
```

创建 localDF 对象并使用 createDataFrame() 函数转换为 Spark DataFrame 对象：

```
> localDF <- data.frame(name=c("John","Mary","Sara"), age=c(19,22,18))
> df <- createDataFrame(localDF)
```

显示对象的结构（Schema）：

```
> printSchema(df) root
```

```
|-- name: string (nullable = true)
|-- age: double (nullable = true)
```

也可使用 str() 函数显示其数据结构:

```
> str(df)
'SparkDataFrame': 2 variables:
$ name: chr "John" "Mary" "Sara"
$ age : num 19 22 18
```

可使用 collect() 函数转换成 R 数据框对象:

```
> local_df <- collect(df)
> str(local_df)
'data.frame': 3 obs. of  2 variables:
$ name: chr  "John" "Mary" "Sara"
$ age : num  19 22 18
```

结束 SparkR:

```
> sparkR.session.stop()
```

注意,结束 SparkR 后,SparkR 的 df 对象已不存在,但 R 的数据框对象还存在。

```
> df
Error in callJMethod(x@sdf, "schema") :
Invalid jobj 15. If SparkR was restarted, Spark operations need to be re- executed.

> local_df
    name age
1   John  19
2   Mary  22
3   Sara  18

> localDF
    name age
1   John  19
2   Mary  22
3   Sara  18
4
```

程序范例 14-3

先使用 RStudio 的 Upload 功能上传 iris.csv(),如图 14-4 所示。

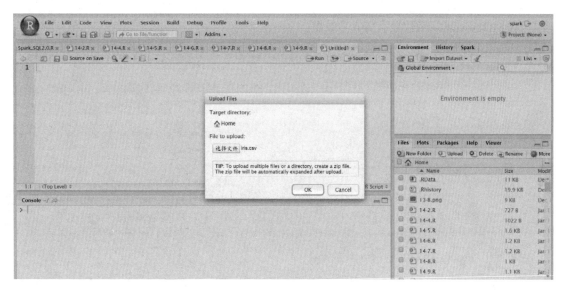

图 14-4　RStudio 的 Upload 功能

设置 SPARK_HOME 路径和使用 SparkR 程序包：

```
> if (nchar(Sys.getenv("SPARK_HOME")) < 1) {
+     Sys.setenv(SPARK_HOME = "/usr/local/spark")
+ }
> library(SparkR, lib.loc = c(file.path(Sys.getenv("SPARK_HOME"), "R", "lib")))
```

启用 sparkR.session 并设置 appName 名称：

```
> sparkR.session(appName = "SparkR-data-manipulation-example")
Java ref type org.apache.spark.sql.SparkSession id 1
```

设置 iris.csv 的路径：

```
> irisCsvPath <- "/home/spark/iris.csv"
```

使用 read.df() 函数读取文件并显示 Schema（注意，此 iris.csv 的 header 与 R 中的 iris.csv 不同，用户应避免使用特殊符号，例如使用"."或空格作为 header）：

```
> irisDF <- read.df(irisCsvPath, source = "csv", header = "true")
> printSchema(irisDF)
root
 |-- _c0: string (nullable = true)
 |-- Sepal_Length: string (nullable = true)
 |-- Sepal_Width: string (nullable = true)
 |-- Petal_Length: string (nullable = true)
 |-- Petal_Width: string (nullable = true)
 |-- Species: string (nullable = true)
```

可将 irisDF 存储到内存中，以提高数据存取的性能：

```
> cache(irisDF)
SparkDataFrame[_c0:string, Sepal_Length:string, Sepal_Width:string, Petal_Length:string, Petal_Width:string, Species:string]
```

可使用 showDF() 函数显示前 6 项数据：

```
> showDF(irisDF, numRows = 6)
+---+------------+-----------+------------+-----------+-------+
|_c0|Sepal_Length|Sepal_Width|Petal_Length|Petal_Width|Species|
+---+------------+-----------+------------+-----------+-------+
|  1|         5.1|        3.5|         1.4|        0.2| setosa|
|  2|         4.9|          3|         1.4|        0.2| setosa|
|  3|         4.7|        3.2|         1.3|        0.2| setosa|
|  4|         4.6|        3.1|         1.5|        0.2| setosa|
|  5|           5|        3.6|         1.4|        0.2| setosa|
|  6|         5.4|        3.9|         1.7|        0.4| setosa|
+---+------------+-----------+------------+-----------+-------+
only showing top 6 rows
```

也可使用 head() 函数显示前 6 项数据：

```
> head(irisDF)
  _c0 Sepal_Length Sepal_Width Petal_Length Petal_Width Species
1   1          5.1         3.5          1.4         0.2  setosa
2   2          4.9           3          1.4         0.2  setosa
3   3          4.7         3.2          1.3         0.2  setosa
4   4          4.6         3.1          1.5         0.2  setosa
5   5            5         3.6          1.4         0.2  setosa
6   6          5.4         3.9          1.7         0.4  setosa
```

可使用 column() 函数显示 header 名称：

```
> columns(irisDF)
[1] "_c0"         "Sepal_Length" "Sepal_Width"  "Petal_Length" "Petal_Width"
[6] "Species"
```

可使用 count() 函数计算总项数：

```
> count(irisDF)
[1] 150
```

选择 Petal_Length 和 Petal_Width 两列数据并显示前 6 项数据：

```
> head(select(irisDF, "Petal_Length", "Petal_Width"))
  Petal_Length Petal_Width
1          1.4         0.2
2          1.4         0.2
3          1.3         0.2
4          1.5         0.2
5          1.4         0.2
6          1.7         0.4
```

使用 where() 函数来筛选出符合 Petal.Length < 1.3 或 Petal.Width < 1 条件的前 6 项数据：

```
head(where(irisDF, irisDF$Petal_Length < 1.3 | irisDF$Petal_Width <1))
  _c0 Sepal_Length Sepal_Width Petal_Length Petal_Width Species
1   1          5.1         3.5          1.4         0.2  setosa
2   2          4.9           3          1.4         0.2  setosa
```

```
3    3          4.7          3.2          1.3          0.2 setosa
4    4          4.6          3.1          1.5          0.2 setosa
5    5          5            3.6          1.4          0.2 setosa
6    6          5.4          3.9          1.7          0.4 setosa
```

使用 arrange() 函数选择 Petal.Length 来进行降序排序并显示前 6 项数据：

```
> head(arrange(irisDF, desc(irisDF$Petal_Length)))
  _c0 Sepal_Length Sepal_Width Petal_Length Petal_Width Species
1 119          7.7         2.6          6.9         2.3 virginica
2 118          7.7         3.8          6.7         2.2 virginica
3 123          7.7         2.8          6.7         2   virginica
4 106          7.6         3            6.6         2.1 virginica
5 132          7.9         3.8          6.4         2   virginica
6 108          7.3         2.9          6.3         1.8 virginica
```

使用 arrange() 函数选择 Petal.Length 进行降序排序、Petal.Width 进行升序排序并显示前 6 项数据：

```
> head(arrange(irisDF, desc(irisDF$Petal_Length), irisDF$Petal_Width))
  _c0 Sepal_Length Sepal_Width Petal_Length Petal_Width Species
1 119          7.7         2.6          6.9         2.3 virginica
2 123          7.7         2.8          6.7         2   virginica
3 118          7.7         3.8          6.7         2.2 virginica
4 106          7.6         3            6.6         2.1 virginica
5 132          7.9         3.8          6.4         2   virginica
6 108          7.3         2.9          6.3         1.8 virginica
```

使用 mutate() 函数产生新的 Petal.Length.new 列数据并显示前 6 项数据：

```
> head(mutate(irisDF, Petal_Length.new = irisDF$Petal_Length/10))
  _c0 Sepal_Length Sepal_Width Petal_Length Petal_Width Species Petal_Length.new
1   1          5.1         3.5          1.4         0.2 setosa              0.14
2   2          4.9         3            1.4         0.2 setosa              0.14
3   3          4.7         3.2          1.3         0.2 setosa              0.13
4   4          4.6         3.1          1.5         0.2 setosa              0.15
5   5          5           3.6          1.4         0.2 setosa              0.14
6   6          5.4         3.9          1.7         0.4 setosa              0.17
```

也可使用 select() 函数来产生新的 Petal.Length.new 行数据并显示前 6 项数据：

```
> head(select(irisDF, irisDF$Petal_Length, irisDF$Petal_Length/10))
  Petal_Length (Petal_Length / 10.0)
1          1.4                  0.14
2          1.4                  0.14
3          1.3                  0.13
4          1.5                  0.15
5          1.4                  0.14
6          1.7                  0.17
```

使用 groupBy()函数并按照 Species 列数据来进行分组，然后使用 count()计算项数并显示前 6 项数据：

```
> head(count(groupBy(irisDF, "Species")))
    Species count
1 virginica    50
2 versicolor   50
3    setosa    50
```

可使用 magrittr 程序包启用管道（Pipe）符号 %>% 功能：

```
> library(magrittr)
```

先使用 groupBy() 函数对 Petal_Width 和 Petal_Length 列数据进行分组，再使用 %>% 传送到 summarize() 函数来计算群组的平均值，最后将结果赋值给 summ_irisDF：

```
> groupBy(irisDF, irisDF$Petal_Width, irisDF$Petal_Length) %>%
+   summarize(avg(irisDF$Petal_Width), avg(irisDF$Petal_Length)) -> summ_irisDF
```

显示前 6 项数据：

```
> head(summ_irisDF)
  Petal_Width Petal_Length avg(Petal_Width) avg(Petal_Length)
1         1.9          5.1              1.9               5.1
2         1.9          5                1.9               5.0
3         2.1          5.9              2.1               5.9
4         2.2          5.8              2.2               5.8
5         1.5          4.2              1.5               4.2
6         1.9          5.3              1.9               5.3
```

可使用 createOrReplaceTempView() 函数将 Spark DataFrame 对象 irisDF 转换为 Spark SQL TABLE（如 irisTable），以便可以使用 Spark SQL 语句：

```
> createOrReplaceTempView(irisDF, "irisTable")
```

其相关的 SQL 语句如下：

```
> sql_selectDF <- sql("SELECT Petal_Length, Petal_Width FROM irisTable")
> head(sql_selectDF)
  Petal_Length Petal_Width
1          1.4         0.2
2          1.4         0.2
3          1.3         0.2
4          1.5         0.2
5          1.4         0.2
6          1.7         0.4

> sql_selectDF <- sql("SELECT * FROM irisTable WHERE Petal_Length <1.3 OR Petal_Width < 1 ")
> head(sql_selectDF)
  _c0 Sepal_Length Sepal_Width Petal_Length Petal_Width Species
1   1          5.1         3.5          1.4         0.2  setosa
2   2          4.9         3            1.4         0.2  setosa
3   3          4.7         3.2          1.3         0.2  setosa
4   4          4.6         3.1          1.5         0.2  setosa
5   5          5            3.6          1.4         0.2  setosa
```

```
6    6         5.4       3.9        1.7       0.4 setosa
> sql_selectDF <- sql("SELECT Species, COUNT(Species) FROM irisTable GROUP BY
Species")
> head(sql_selectDF)
     Species count(Species)
1  virginica             50
2 versicolor             50
3     setosa             50

> sql_selectDF <- sql("SELECT Petal_Width, Petal_Length FROM irisTable ORDER
BY Petal_Width ASC")
> head(sql_selectDF)
  Petal_Width Petal_Length
1         0.1          1.4
2         0.1          1.5
3         0.1          1.5
4         0.1          1.4
5         0.1          1.1
6         0.2          1.5
> sql_selectDF <- sql("SELECT Petal_Width, Petal_Length FROM irisTable ORDER
BY Petal_Width DESC")
> head(sql_selectDF)
  Petal_Width Petal_Length
1         2.5          6
2         2.5          6.1
3         2.5          5.7
4         2.4          5.1
5         2.4          5.6
6         2.4          5.6
```

结束 SparkR：

```
> sparkR.session.stop()
```

14.3　SparkR 与 SQL Server

读者可先将本书的 IRIS_Data 数据库附加到 MS SQL Server 中（注意要将防火墙 port1433 打开，并在 SQL Server 配置管理器中启用 TCP/IP）。接着下载 Microsoft JDBC Driver 4.0 for SQL Server（https://www.microsoft.com/zh-cn/download/details.aspx?id=11774）到 /usr/local/spark 目录下。

程序范例 14-4

首先设置 SPARK_HOME 路径和使用 SparkR 程序包：

```
> if (nchar(Sys.getenv("SPARK_HOME")) < 1) {
+     Sys.setenv(SPARK_HOME = "/usr/local/spark")
+ }
```

```
> library(SparkR, lib.loc = c(file.path(Sys.getenv("SPARK_HOME"), "R", "lib")))
```

设置 JDBC Driver：

```
> Sys.setenv(SPARK_CLASSPATH="/usr/local/spark/sqljdbc4.jar")
```

启用 sparkR.session 并设置 appName 名称：

```
> sparkR.session(appName = "SparkR-SQL Server")
Java ref type org.apache.spark.sql.SparkSession id 1
```

使用 read.jdbc 连接 sqlserver 并设置存取 iris 表格、账号及密码：

```
>       df<-read.jdbc("jdbc:sqlserver://      192.168.28.1",tableName="iris",user="test",password="test")
```

显示前 6 项数据：

```
> head(df)
  sepal_length sepal_width petal_length petal_width species
1          5.1         3.5          1.4         0.2  setosa
2          4.9           3          1.4         0.2  setosa
3          4.7         3.2          1.3         0.2  setosa
4          4.6         3.1          1.5         0.2  setosa
5            5         3.6          1.4         0.2  setosa
6          5.4         3.9          1.7         0.4  setosa
```

计算总项数：

```
> count(df)
[1] 150
```

也可使用 createOrReplaceTempView() 函数将 Spark DataFrame 对象 df 转换为 Spark SQL TABLE（如 irisTable），以便使用程序范例 14-3 的 Spark SQL 相关语句：

```
> createOrReplaceTempView(df, "irisTable")

> sql_selectDF <- sql("SELECT Petal_Length, Petal_Width FROM irisTable")
> sql_selectDF
SparkDataFrame[Petal_Length:string, Petal_Width:string]

> head(sql_selectDF)
  Petal_Length Petal_Width
1          1.4         0.2
2          1.4         0.2
3          1.3         0.2
4          1.5         0.2
5          1.4         0.2
6          1.7         0.4
```

用户可使用 ggplot2 程序包的 qplot() 函数来显示图形：

```
> library(ggplot2)
```

先使用 collect() 函数将 df 转换成 R 数据框对象，以 sepal_length 为 X 轴、sepal_width 为 Y 轴并以颜色显示 species：

```
> qplot(data=collect(df), sepal_length, sepal_width, colour=species)
```

其结果如图 14-5 所示。

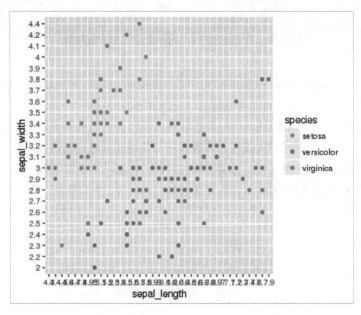

图 14-5　以 qplot 显示 species 颜色

以 sepal_length 为 X 轴、sepal_width 为 Y 轴并以形状显示 species。

```
> qplot(data=collect(df), sepal_length, sepal_width, shape=species)
```

其结果如图 14-6 所示。

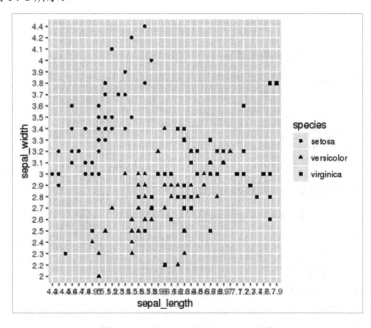

图 14-6　以 qplot 显示 species 形状

结束 SparkR：

```
> sparkR.session.stop()
```

14.4　SparkR 与 Cassandra

Cassandra 是采用 Key-Value 数据结构的分布式数据库，2007 年由 Facebook 开发，2009 年成为 Apache 的孵化项目。Cassandra 取消了原本关系数据库中常用的 Schema 结构，可以打造出分布式和高扩充能力的特性。Cassandra 的数据可写入多个节点，来保证数据的可靠性、一致性、可用性，其适用于可能有节点网络失效以及多数据中心的场景；可无缝地加入或删除节点，非常适合节点规模变化比较快的应用领域。在新版本的 Cassandra 中开发了 CQL（Cassandra Query Language），其功能和 SQL 的作用类似，可用于操作 Cassandra 中的数据。用户可参考附录 F 安装相关的 Driver、启动 Cassandra 以及创建执行范例所需的 Keyspace 和 Table。

程序范例 14-5

使用 RJDBC 连接 Cassandra 及查询数据，首先使用 RJDBC 程序包：

```
> library(RJDBC)
```

启动 JDBC Driver：

```
> drv <- JDBC("org.apache.cassandra.cql.jdbc.CassandraDriver",list.files("/usr/local/apache-cassandra-3.7/lib",pattern = "jar$", full.names = TRUE))
```

使用 dbConnect() 函数连接已建立于 192.168.244.150:9160/mykeyspace 中的 Cassandra Keyspace：

```
> conn <- dbConnect(drv, "jdbc:cassandra://192.168.244.150:9160/mykeyspace")
```

使用 CQL 指令查询 student 数据表：

```
> res <- dbGetQuery(conn, "select * from student")
> head(res)
         no class  name
1  B0001006  IM1A Frodo
2  B0001007  IM2A Brodo
3  B0001008  IM3A Krodo
```

使用 CQL 指令查询 student 数据表的数据项数：

```
> res <- dbGetQuery(conn, "select count(*) from student")
> res
  count
1     3
```

用户可使用 SparkR 来读取 Cassandra 数据，使用方式如下。

程序范例 14-6

首先设置 SPARK_HOME 路径和使用 SparkR 程序包:

```
> if (nchar(Sys.getenv("SPARK_HOME")) < 1) {
+ Sys.setenv(SPARK_HOME = "/usr/local/spark")
+ }
> library(SparkR, lib.loc = c(file.path(Sys.getenv("SPARK_HOME"), "R", "lib")))
```

设置 spark-cassandra-connector-2.0.0-M2-s_2.11.jar 环境变量:

```
> Sys.setenv('SPARKR_SUBMIT_ARGS'='"--packages" "com.datastax.spark:spark-cassandra-connector:2.0.0-M2-s_2.11" "--conf" "spark.cassandra.connection.host=192.168.244.150" "--jars" "/usr/local/spark/spark-cassandra-connector- 2.0.0-M2-s_2.11.jar" "sparkr-shell"')
```

启用 sparkR.session 并设置 appName 名称:

```
> sparkR.session(appName = "SparkR-Cassandra")
Java ref type org.apache.spark.sql.SparkSession id 1
```

使用 read.df() 函数读取 mykeyspace 的 student 数据表并赋值给 people 对象,调用 head() 函数来显示 people 对象的数据:

```
> people <-read.df("192.168.244.150", source = "org.apache.spark.sql. cassandra", keyspace = "mykeyspace", table = "student")
> head(people)
         no  class  name
1  B0001007  IM2A   Brodo
2  B0001008  IM3A   Krodo
3  B0001006  IM1A   Frodo
```

确定 people 对象为 Spark DataFrame:

```
> str(people) 'SparkDataFrame': 3 variables:
$ no    : chr "B0001007" "B0001008" "B0001006"
$ class : chr "IM2A" "IM3A" "IM1A"
$ name  : chr "Brodo" "Krodo" "Frodo"
```

使用 createOrReplaceTempView() 函数把 people 对象转换为 Spark SQL TABLE,如 peopleTable:

```
> createOrReplaceTempView(people, "peopleTable")
```

可使用程序范例 14-3 的 Spark SQL 相关语句并调用 head()函数显示数据:

```
> sql_selectDF <- sql("SELECT no, class FROM peopleTable")
> head(sql_selectDF)
        no  class
1 B0001007  IM2A
2 B0001008  IM3A
3 B0001006  IM1A
```

结束 SparkR:

```
> sparkR.session.stop()
```

14.5 Spark Standalone 模式

用户可参考附录 F 建立 Spark Standalone 模式。启动时只需在 192.168.244.151 (master) 上执行：

```
# su - spark
$ cd /usr/local/spark
$ sbin/start-all.sh
```

结束 Spark Standlone：

```
$ sbin/stop-all.sh
```

用户可使用 192.168.244.151:8080 确认 master 和 worker 的执行结果。

程序范例 14-7

首先清除所有对象：

```
> rm(list = ls())
> gc()
         used  (Mb) gc    trigger (Mb) max used (Mb)
Ncells  454637  24.3        940480 50.3   750400  40.1
Vcells  681620   5.3      12533012 95.7 18937484 144.5
```

设置 SPARK_HOME 路径和使用 SparkR 程序包：

```
> if (nchar(Sys.getenv("SPARK_HOME")) < 1) {
+ Sys.setenv(SPARK_HOME = "/usr/local/spark")
+ }
> library(SparkR, lib.loc = c(file.path(Sys.getenv("SPARK_HOME"), "R", "lib")))
```

启用 sparkR.session 并设置 spark.driver.memory 为 20G（用户可按照自己的系统设置此参数）：

```
> sparkR.session(master = "spark://192.168.244.151:7077", sparkConfig = 
list(spark.driver.memory = "20g"))
  Java ref type org.apache.spark.sql.SparkSession id 1
```

设置 allyears_10.csv 路径（用户可到 https://drive.google.com/file/d/0BytXz-NUNBibMzdJSkdfRjJlRW8/view?usp=sharing 下载这个 31G 的文件）：

```
> allyearsCsvPath <- "/home/spark/allyears_10.csv"
```

使用 proc.time() 和 read.df() 函数计算读文件的时间：

```
> ptm <- proc.time()
> allyearsDF <- read.df(allyearsCsvPath, source = "csv", header = "true")
> proc.time()-ptm
  user  system elapsed
```

```
 0.018    0.007    6.868
```

使用 createOrReplaceTempView() 函数将 Spark DataFrame 对象 allyearsDF 转换为 Spark SQL TABLE（如 allyearsTable），以便可以使用 Spark SQL 语句：

```
> createOrReplaceTempView(allyearsDF, "allyearsTable")
```

使用 sql() 函数读取所有数据、显示前 6 项数据并计算运行时间：

```
> ptm <- proc.time()
> sql_selectDF <- sql("SELECT * FROM allyearsTable")
> head(sql_selectDF)
17/01/05 22:22:43 WARN Utils: Truncated the string representation of a plan since it was too
large. This behavior can be adjusted by setting 'spark.debug.maxToStringFields' in SparkEnv.conf.
  Year Month DayofMonth DayOfWeek DepTime CRSDepTime ArrTime CRSArrTime UniqueCarrier FlightNum
1 1987    10         14         3     741        730     912        849            PS      1451
2 1987    10         15         4     729        730     903        849            PS      1451
3 1987    10         17         6     741        730     918        849            PS      1451
4 1987    10         18         7     729        730     847        849            PS      1451
5 1987    10         19         1     749        730     922        849            PS      1451
6 1987    10         21         3     728        730     848        849            PS      1451
  TailNum ActualElapsedTime CRSElapsedTime AirTime ArrDelay DepDelay Origin Dest Distance TaxiIn
1      NA                91             79      NA       23       11    SAN  SFO      447     NA
2      NA                94             79      NA       14       -1    SAN  SFO      447     NA
3      NA                97             79      NA       29       11    SAN  SFO      447     NA
4      NA                78             79      NA       -2       -1    SAN  SFO      447     NA
5      NA                93             79      NA       33       19    SAN  SFO      447     NA
6      NA                80             79      NA       -1       -2    SAN  SFO      447     NA
  TaxiOut Cancelled CancellationCode Diverted CarrierDelay WeatherDelay NASDelay SecurityDelay
1      NA         0               NA        0           NA           NA       NA            NA
2      NA         0               NA        0           NA           NA       NA            NA
3      NA         0               NA        0           NA           NA       NA            NA
4      NA         0               NA        0           NA           NA       NA            NA
5      NA         0               NA        0           NA           NA       NA            NA
6      NA         0               NA        0           NA           NA       NA            NA
  LateAircraftDelay IsArrDelayed IsDepDelayed
1                NA          YES          YES
2                NA          YES           NO
3                NA          YES          YES
4                NA           NO           NO
5                NA          YES          YES
6                NA           NO           NO
> proc.time()-ptm
   user  system elapsed
  0.113   0.026   3.516
```

使用 sql() 函数读取 ArrDelay <= 0 AND Distance=337 的 Year、DayofMonth、DayOfWeek、ArrTime、DepDelay、Origin、Dest 等字段的数据、显示前 6 项数据并计算运行时间：

```
> ptm <- proc.time()
> new_dfx1 <- sql("SELECT Year, DayofMonth, DayOfWeek, ArrTime, DepDelay,
  Origin, Dest FROM allyearsTable WHERE ArrDelay <= 0 AND Distance = 337")
```

```
> head(new_dfx1)
  Year DayofMonth DayOfWeek ArrTime DepDelay Origin Dest
1 1987          5         1     932        3    LAX  SFO
2 1987         26         1     933        1    LAX  SFO
3 1987         27         2     936        2    LAX  SFO
4 1987          5         1    1032        1    LAX  SFO
5 1987         10         6    1032        0    LAX  SFO
6 1987         17         6    1035       -1    LAX  SFO
> proc.time()-ptm
   user  system elapsed
  0.083   0.012   2.943
```

可使用 collect() 函数转换成 R 数据框对象并计算运行时间：

```
> ptm <- proc.time()
> dfy1 <- collect(new_dfx1)
[Stage 4:==============================================
=========>(246 + 2) / 248][Stage 4:==================
===================================>(247 + 1) / 248]
> proc.time()-ptm
   user  system elapsed
324.318   0.876 543.829
```

结束 SparkR：

```
> sparkR.session.stop()
```

由以上可知，Standalone 模式调用 collect() 函数执行的 elapsed 时间为 543.829 秒。用户可进一步比较 Standalone 和 Local 模式下的执行性能。

程序范例 14-8

相同的程序代码再以 Local 模式执行：

```
> if (nchar(Sys.getenv("SPARK_HOME")) < 1) {
+     Sys.setenv(SPARK_HOME = "/usr/local/spark")
+ }
> library(SparkR, lib.loc = c(file.path(Sys.getenv("SPARK_HOME"), "R", "lib")))
```

设置为 Local 模式：

```
> sparkR.session(master = "local", sparkConfig = list(spark.driver.memory = "20g"))
> allyearsDF <- read.df(allyearsCsvPath, source = "csv", header = "true")
> createOrReplaceTempView(allyearsDF, "allyearsTable")
> sql_selectDF <- sql("SELECT * FROM allyearsTable")
> head(sql_selectDF)
> new_dfx1 <- sql("SELECT Year, DayofMonth, DayOfWeek, ArrTime, DepDelay,
Origin, Dest FROM allyearsTable
+               WHERE ArrDelay <= 0 AND Distance = 337")

> ptm <- proc.time()
> dfy1 <- collect(new_dfx1)
```

```
> proc.time()-ptm
   user  system elapsed
295.855   1.883 1507.366
```

由以上可知，Local 模式调用 collect() 函数执行的 elapsed 时间为 1507.366 秒。用户可调用 qplot() 函数来显示如图 14-7 所示的图标信息：

```
> qplot(data=dfy1, x=Dest, y=DepDelay, color= Dest)
```

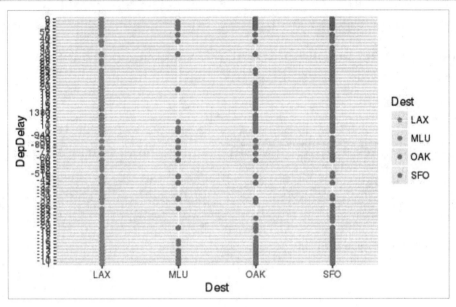

图 14-7　以 qplot 显示 dfy1 信息

结束 SparkR：

```
> sparkR.session.stop()
```

14.6　SparkR 数据分析

用户也可以使用 SparkR 提供的 API 工具来分析数据，其网址为 http://spark.apache.org/docs/latest/api/R/index.html。

程序范例 14-9

首先清除所有对象：

```
> rm(list = ls())
> gc()
```

设置 SPARK_HOME 路径和使用 SparkR 程序包：

```
> if (nchar(Sys.getenv("SPARK_HOME")) < 1) {
+ Sys.setenv(SPARK_HOME = "/usr/local/spark")
```

```
+ }
> library(SparkR, lib.loc = c(file.path(Sys.getenv("SPARK_HOME"), "R", "lib")))
```

启用 sparkR.session 并设置 appName 名称：

```
> sparkR.session(appName = " SparkR-data-analysis-example")
Java ref type org.apache.spark.sql.SparkSession id 1
```

使用 createDataFrame()函数将 iris 数据集转换为 Spark DataFrame 对象且不显示警告信息，将此对象赋值给 irisDF 对象：

```
> irisDF <- suppressWarnings(createDataFrame(iris))
```

将 irisDF 对象赋值给 kmeansDF 对象并当作执行 spark.kmeans() 函数的输入数据，k=3 表示分为 3 个分组。注意 kmeans 并无输出变量。

```
> kmeansDF <- irisDF
> kmeansModel <- spark.kmeans(kmeansDF, ~ Sepal_Length + Sepal_Width + Petal_Length + Petal_Width,k = 3)
17/01/17 22:14:08 WARN KMeans: The input data is not directly cached, which may hurt performance if its parent RDDs are also uncached.
17/01/17 22:14:10 WARN KMeans: The input data was not directly cached, which may hurt performance if its parent RDDs are also uncached.
[Stage 146:================================================>  (185 + 1) / 200]
```

使用 summary() 函数显示信息：

```
> summary(kmeansModel)
$k
[1] 3

$coefficients
  Sepal_Length Sepal_Width Petal_Length Petal_Width
1 5.883607     2.740984    4.388525     1.434426
2 5.006        3.428       1.462        0.246
3 6.853846     3.076923    5.715385     2.053846

$size
$size[[1]]
[1] 61

$size[[2]]
[1] 50

$size[[3]]
[1] 39


$cluster
SparkDataFrame[prediction:int]

$is.loaded
[1] FALSE
```

SparkR 大数据分析 第 14 章

使用 fitted()和 head()函数来显示前 6 项数据:

```
> fitted_kmeans=fitted(kmeansModel)
> head(fitted_kmeans)
  Sepal_Length Sepal_Width Petal_Length Petal_Width Species prediction
1          5.1         3.5          1.4         0.2  setosa          1
2          4.9         3.0          1.4         0.2  setosa          1
3          4.7         3.2          1.3         0.2  setosa          1
4          4.6         3.1          1.5         0.2  setosa          1
5          5.0         3.6          1.4         0.2  setosa          1
6          5.4         3.9          1.7         0.4  setosa          1
```

接着使用 spark.randomForest() 函数来给 iris 数据集分类。使用 kmeansDF 作为训练集数据、Species 作为输出变量来存放分类结果,度用 10 棵决策树来建立随机森林:

```
> model <- spark.randomForest(kmeansDF, Species ~ ., "classification", numTrees = 10)
```

使用 summary() 函数来显示决策规则:

```
> summary(model)
Formula: Species ~ .
Number of features: 4
Features: Sepal_Length Sepal_Width Petal_Length Petal_Width
Feature importances: (4,[0,1,2,3],[0.06895553555421127,0.010289460177385103,0
.3240437199870398,0.5967112842813639])
Number of trees: 10
Tree weights: 1 1 1 1 1 1 1 1 1 1
RandomForestClassificationModel (uid=rfc_7c6536c556ba) with 10 trees
  Tree 0 (weight 1.0):
    If (feature 3 <= 0.6)
     Predict: 2.0
    Else (feature 3 > 0.6)
     If (feature 3 <= 1.6)
      If (feature 0 <= 6.8)
       If (feature 1 <= 2.6)
        If (feature 0 <= 6.0)
         Predict: 0.0
        Else (feature 0 > 6.0)
         Predict: 0.0
       Else (feature 1 > 2.6)
        Predict: 0.0
      Else (feature 0 > 6.8)
       Predict: 1.0
     Else (feature 3 > 1.6)
      If (feature 3 <= 1.7)
       If (feature 1 <= 2.5)
        Predict: 1.0
       Else (feature 1 > 2.5)
        Predict: 0.0
      Else (feature 3 > 1.7)
       If (feature 0 <= 5.9)
```

```
            If (feature 1 <= 3.0)
              Predict: 1.0
            Else (feature 1 > 3.0)
              Predict: 0.0
         Else (feature 0 > 5.9)
           Predict: 1.0
  Tree 1 (weight 1.0):
     If (feature 3 <= 0.6)
       Predict: 2.0
     Else (feature 3 > 0.6)
       If (feature 3 <= 1.7)
         If (feature 2 <= 5.0)
           Predict: 0.0
         Else (feature 2 > 5.0)
           Predict: 1.0
       Else (feature 3 > 1.7)
         If (feature 0 <= 5.9)
           If (feature 0 <= 5.8)
             Predict: 1.0
           Else (feature 0 > 5.8)
             If (feature 2 <= 4.8)
               Predict: 0.0
             Else (feature 2 > 4.8)
               Predict: 1.0
         Else (feature 0 > 5.9)
           Predict: 1.0
  Tree 2 (weight 1.0):
     If (feature 3 <= 1.7)
       If (feature 3 <= 0.4)
         Predict: 2.0
       Else (feature 3 > 0.4)
         If (feature 2 <= 5.1)
           If (feature 2 <= 4.9)
             Predict: 0.0
           Else (feature 2 > 4.9)
             If (feature 1 <= 2.2)
               Predict: 1.0
             Else (feature 1 > 2.2)
               Predict: 0.0
         Else (feature 2 > 5.1)
           Predict: 1.0
     Else (feature 3 > 1.7)
       Predict: 1.0
  Tree 3 (weight 1.0):
     If (feature 3 <= 0.6)
       Predict: 2.0
     Else (feature 3 > 0.6)
       If (feature 3 <= 1.6)
         If (feature 0 <= 6.0)
           Predict: 0.0
```

```
    Else (feature 0 > 6.0)
     If (feature 0 <= 6.1)
      If (feature 2 <= 4.6)
       Predict: 0.0
      Else (feature 2 > 4.6)
       Predict: 1.0
     Else (feature 0 > 6.1)
      If (feature 3 <= 1.4)
       Predict: 0.0
      Else (feature 3 > 1.4)
       Predict: 0.0
   Else (feature 3 > 1.6)
    If (feature 2 <= 5.0)
     If (feature 1 <= 3.0)
      If (feature 0 <= 6.3)
       Predict: 1.0
      Else (feature 0 > 6.3)
       Predict: 0.0
     Else (feature 1 > 3.0)
      Predict: 0.0
    Else (feature 2 > 5.0)
     Predict: 1.0
  Tree 4 (weight 1.0):
    If (feature 2 <= 1.9)
     Predict: 2.0
    Else (feature 2 > 1.9)
     If (feature 2 <= 4.7)
      If (feature 2 <= 4.4)
       Predict: 0.0
      Else (feature 2 > 4.4)
       If (feature 3 <= 1.5)
        Predict: 0.0
       Else (feature 3 > 1.5)
        Predict: 1.0
     Else (feature 2 > 4.7)
      If (feature 2 <= 4.9)
       If (feature 0 <= 6.1)
        Predict: 1.0
       Else (feature 0 > 6.1)
        If (feature 3 <= 1.5)
         Predict: 0.0
        Else (feature 3 > 1.5)
         Predict: 1.0
      Else (feature 2 > 4.9)
       If (feature 3 <= 1.6)
        If (feature 0 <= 6.0)
         Predict: 0.0
        Else (feature 0 > 6.0)
         Predict: 1.0
       Else (feature 3 > 1.6)
```

```
      Predict: 1.0
Tree 5 (weight 1.0):
   If (feature 2 <= 1.9)
    Predict: 2.0
   Else (feature 2 > 1.9)
    If (feature 2 <= 4.7)
      If (feature 1 <= 2.5)
        If (feature 0 <= 4.9)
         Predict: 1.0
        Else (feature 0 > 4.9)
         Predict: 0.0
      Else (feature 1 > 2.5)
        Predict: 0.0
    Else (feature 2 > 4.7)
      If (feature 3 <= 1.6)
        If (feature 2 <= 4.9)
         Predict: 0.0
        Else (feature 2 > 4.9)
         If (feature 3 <= 1.5)
           Predict: 1.0
         Else (feature 3 > 1.5)
           Predict: 0.0
      Else (feature 3 > 1.6)
        Predict: 1.0
Tree 6 (weight 1.0):
   If (feature 3 <= 0.4)
    Predict: 2.0
   Else (feature 3 > 0.4)
    If (feature 3 <= 1.7)
      If (feature 2 <= 5.1)
        If (feature 2 <= 4.9)
         If (feature 3 <= 1.6)
           Predict: 0.0
         Else (feature 3 > 1.6)
           Predict: 1.0
        Else (feature 2 > 4.9)
         If (feature 3 <= 1.5)
           Predict: 1.0
         Else (feature 3 > 1.5)
           Predict: 0.0
      Else (feature 2 > 5.1)
        Predict: 1.0
    Else (feature 3 > 1.7)
      If (feature 2 <= 4.8)
        If (feature 0 <= 5.9)
         Predict: 0.0
        Else (feature 0 > 5.9)
         Predict: 1.0
      Else (feature 2 > 4.8)
        Predict: 1.0
```

```
Tree 7 (weight 1.0):
  If (feature 0 <= 5.5)
   If (feature 2 <= 1.9)
    Predict: 2.0
   Else (feature 2 > 1.9)
    Predict: 0.0
  Else (feature 0 > 5.5)
   If (feature 2 <= 5.0)
    If (feature 3 <= 0.4)
     Predict: 2.0
    Else (feature 3 > 0.4)
     If (feature 2 <= 4.7)
      Predict: 0.0
     Else (feature 2 > 4.7)
      If (feature 3 <= 1.7)
       Predict: 0.0
      Else (feature 3 > 1.7)
       Predict: 1.0
   Else (feature 2 > 5.0)
    If (feature 3 <= 1.6)
     If (feature 1 <= 2.7)
      Predict: 0.0
     Else (feature 1 > 2.7)
      Predict: 1.0
    Else (feature 3 > 1.6)
     Predict: 1.0
Tree 8 (weight 1.0):
  If (feature 3 <= 0.5)
   Predict: 2.0
  Else (feature 3 > 0.5)
   If (feature 3 <= 1.7)
    If (feature 3 <= 1.3)
     Predict: 0.0
    Else (feature 3 > 1.3)
     If (feature 2 <= 5.1)
      If (feature 0 <= 4.9)
       Predict: 1.0
      Else (feature 0 > 4.9)
       Predict: 0.0
     Else (feature 2 > 5.1)
      Predict: 1.0
   Else (feature 3 > 1.7)
    If (feature 2 <= 4.8)
     If (feature 0 <= 5.9)
      Predict: 0.0
     Else (feature 0 > 5.9)
      Predict: 1.0
    Else (feature 2 > 4.8)
     Predict: 1.0
  Tree 9 (weight 1.0):
```

```
          If (feature 2 <= 1.9)
           Predict: 2.0
          Else (feature 2 > 1.9)
           If (feature 0 <= 6.7)
            If (feature 3 <= 1.6)
             If (feature 2 <= 4.9)
              Predict: 0.0
             Else (feature 2 > 4.9)
              If (feature 0 <= 6.0)
               Predict: 0.0
              Else (feature 0 > 6.0)
               Predict: 1.0
            Else (feature 3 > 1.6)
             If (feature 1 <= 3.1)
              Predict: 1.0
             Else (feature 1 > 3.1)
              If (feature 0 <= 5.9)
               Predict: 0.0
              Else (feature 0 > 5.9)
               Predict: 1.0
           Else (feature 0 > 6.7)
            If (feature 3 <= 1.4)
             Predict: 0.0
            Else (feature 3 > 1.4)
             Predict: 1.0
```

使用 predict() 函数来预测 kmeansDF 数据集并使用 showDF() 函数来显示数据：

```
> predictions <- predict(model, kmeansDF)
> showDF(predictions,6)
+-----------+----------+------------+----------+------+--------------+--------------+----------+
|Sepal_Length|Sepal_Width|Petal_Length|Petal_Width|Species|  rawPrediction|   probability|prediction|
+-----------+----------+------------+----------+------+--------------+--------------+----------+
|        5.1|       3.5|         1.4|       0.2|setosa|[0.0,0.0,10.0]|[0.0,0.0,1.0]|    setosa|
|        4.9|       3.0|         1.4|       0.2|setosa|[0.0,0.0,10.0]|[0.0,0.0,1.0]|    setosa|
|        4.7|       3.2|         1.3|       0.2|setosa|[0.0,0.0,10.0]|[0.0,0.0,1.0]|    setosa|
|        4.6|       3.1|         1.5|       0.2|setosa|[0.0,0.0,10.0]|[0.0,0.0,1.0]|    setosa|
|        5.0|       3.6|         1.4|       0.2|setosa|[0.0,0.0,10.0]|[0.0,0.0,1.0]|    setosa|
|        5.4|       3.9|         1.7|       0.4|setosa|[0.0,0.0,10.0]|[0.0,0.0,1.0]|    setosa|
+-----------+----------+------------+----------+------+--------------+--------------+----------+
only showing top 6 rows
```

结束 SparkR：

```
> sparkR.session.stop()
```

附录 A 下载和安装 R

下载 R（读者可参考下列步骤安装适合的 R 版本）

步骤 01 R 网站位于 http://www.r-project.org/，网站首页如图 A-1 所示。

图 A-1

步骤 02 选择下载网站，如图 A-2 所示。

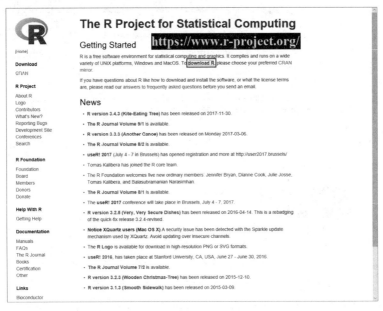

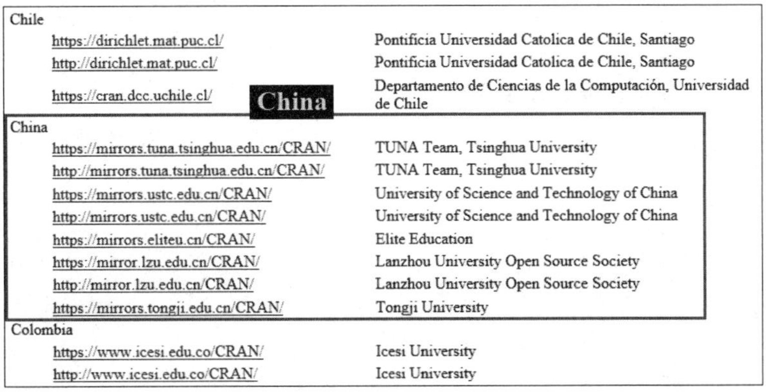

Chile	
https://dirichlet.mat.puc.cl/	Pontificia Universidad Catolica de Chile, Santiago
http://dirichlet.mat.puc.cl/	Pontificia Universidad Catolica de Chile, Santiago
https://cran.dcc.uchile.cl/	Departamento de Ciencias de la Computación, Universidad de Chile
China	
https://mirrors.tuna.tsinghua.edu.cn/CRAN/	TUNA Team, Tsinghua University
http://mirrors.tuna.tsinghua.edu.cn/CRAN/	TUNA Team, Tsinghua University
https://mirrors.ustc.edu.cn/CRAN/	University of Science and Technology of China
http://mirrors.ustc.edu.cn/CRAN/	University of Science and Technology of China
https://mirrors.eliteu.cn/CRAN/	Elite Education
https://mirror.lzu.edu.cn/CRAN/	Lanzhou University Open Source Society
http://mirror.lzu.edu.cn/CRAN/	Lanzhou University Open Source Society
https://mirrors.tongji.edu.cn/CRAN/	Tongji University
Colombia	
https://www.icesi.edu.co/CRAN/	Icesi University
http://www.icesi.edu.co/CRAN/	Icesi University

图 A-2

步骤 03 选择操作系统，如图 A-3 所示。

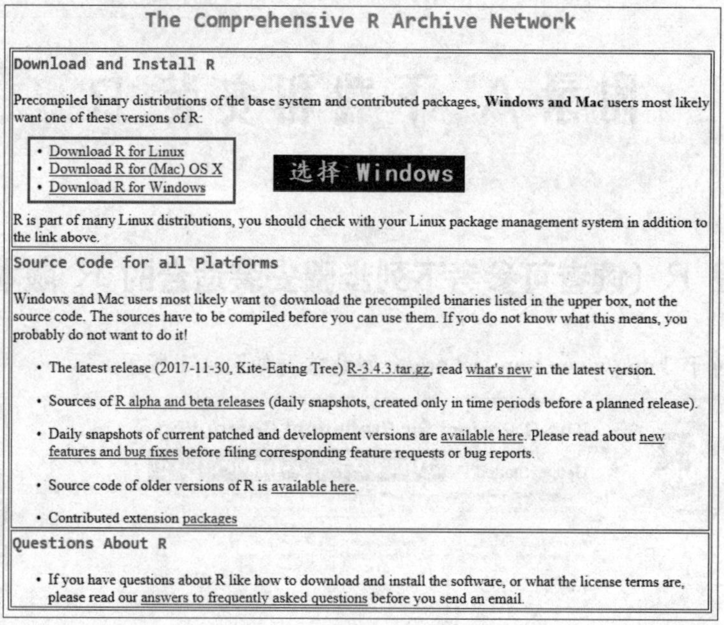

The Comprehensive R Archive Network

Download and Install R

Precompiled binary distributions of the base system and contributed packages, **Windows and Mac** users most likely want one of these versions of R:

- Download R for Linux
- Download R for (Mac) OS X
- Download R for Windows

选择 Windows

R is part of many Linux distributions, you should check with your Linux package management system in addition to the link above.

Source Code for all Platforms

Windows and Mac users most likely want to download the precompiled binaries listed in the upper box, not the source code. The sources have to be compiled before you can use them. If you do not know what this means, you probably do not want to do it!

- The latest release (2017-11-30, Kite-Eating Tree) R-3.4.3.tar.gz, read what's new in the latest version.
- Sources of R alpha and beta releases (daily snapshots, created only in time periods before a planned release).
- Daily snapshots of current patched and development versions are available here. Please read about new features and bug fixes before filing corresponding feature requests or bug reports.
- Source code of older versions of R is available here.
- Contributed extension packages

Questions About R

- If you have questions about R like how to download and install the software, or what the license terms are, please read our answers to frequently asked questions before you send an email.

图 A-3

步骤 04 选择 base 下载，如图 A-4 所示。

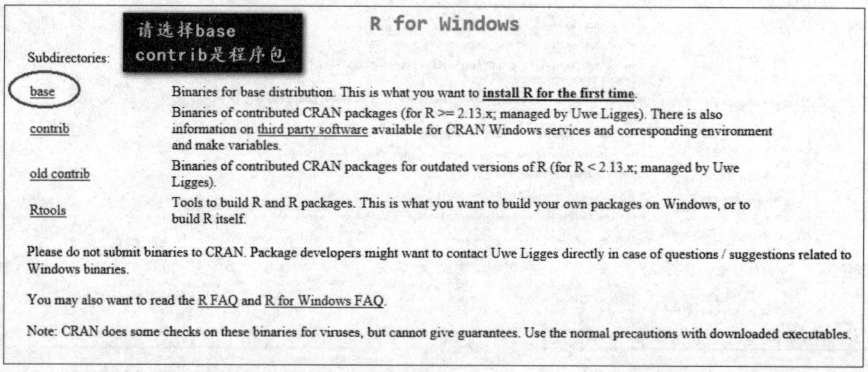

R for Windows

请选择base
contrib是程序包

Subdirectories:

base	Binaries for base distribution. This is what you want to **install R for the first time**.
contrib	Binaries of contributed CRAN packages (for R >= 2.13.x; managed by Uwe Ligges). There is also information on third party software available for CRAN Windows services and corresponding environment and make variables.
old contrib	Binaries of contributed CRAN packages for outdated versions of R (for R < 2.13.x; managed by Uwe Ligges).
Rtools	Tools to build R and R packages. This is what you want to build your own packages on Windows, or to build R itself.

Please do not submit binaries to CRAN. Package developers might want to contact Uwe Ligges directly in case of questions / suggestions related to Windows binaries.

You may also want to read the R FAQ and R for Windows FAQ.

Note: CRAN does some checks on these binaries for viruses, but cannot give guarantees. Use the normal precautions with downloaded executables.

图 A-4

步骤 05 选择版本，如图 A-5 和图 A-6 所示。

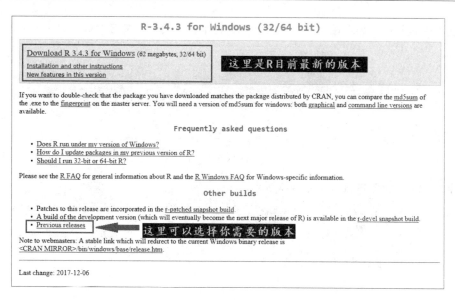

图 A-5

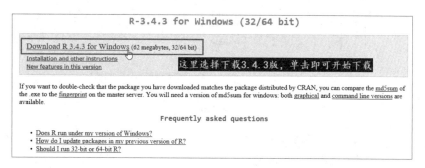

图 A-6

安装步骤

步骤 01 在下载文件夹中看到 "R-3.4.3-win.exe" 安装文件，如图 A-7 所示。

图 A-7

步骤02 执行安装程序，先选择"中文（简体）"再单击"确定"按钮，如图A-8所示。

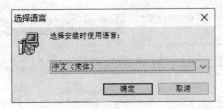

图 A-8

步骤03 单击"下一步"按钮，如图A-9所示。

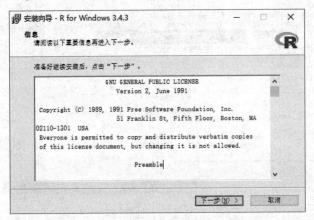

图 A-9

步骤04 单击"浏览"按钮，选择所要安装的文件夹，再单击"下一步"按钮，如图A-10所示。

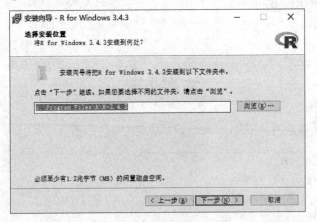

图 A-10

步骤05 选择所要安装的组件，再单击"下一步"按钮，如图A-11所示。

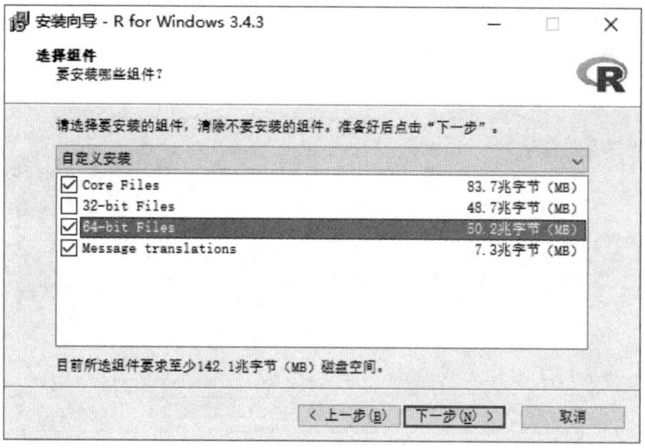

图 A-11

步骤 06 保留默认选择"No（接受默认选项）"，再单击"下一步"按钮，如图 A-12 所示。

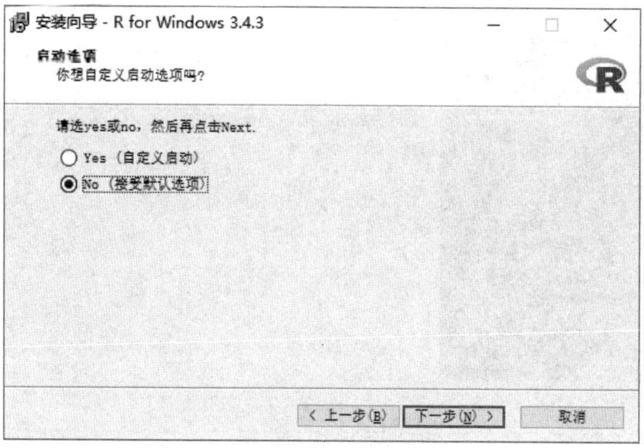

图 A-12

步骤 07 确认无误后，单击"下一步"按钮，如图 A-13 所示。

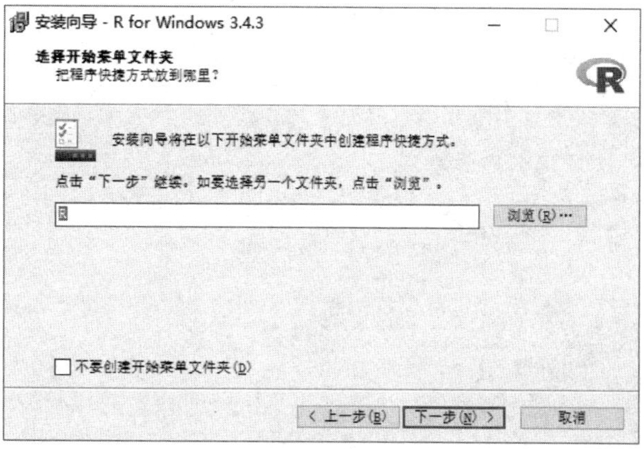

图 A-13

步骤 08 选择附加任务所需的选项，再单击"下一步"按钮，如图 A-14 所示。

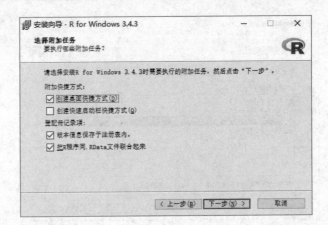

图 A-14

步骤 09 单击"结束"按钮,即可完成 R 软件的安装,如图 A-15 所示。

图 A-15

附录 B 安装 RStudio Desktop

步骤01 前往网站 https://www.rstudio.com/ 下载 RStudio Desktop，如图 B-1 所示。

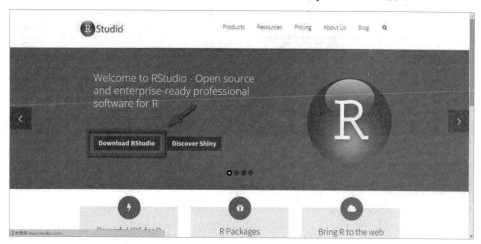

图 B-1

步骤02 选择"Desktop"，如图 B-2 所示。出现如图 B-3 所示的网页后，单击"DOWNLOAD RSTUDIO DESKTOP"按钮。再到显示出的如图 B-4 所示的网页中选择操作系统，选择完之后就会开始下载 RStudio 安装程序文件。

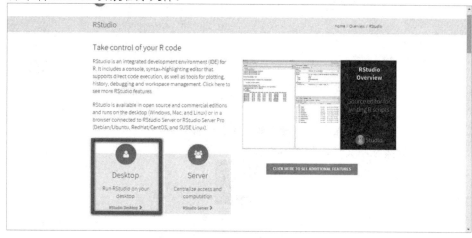

图 B-2

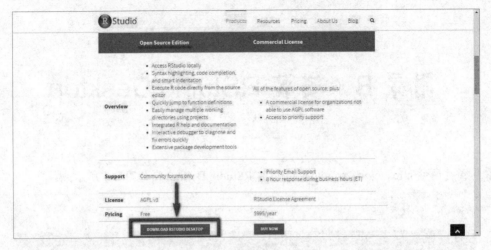

图 B-3

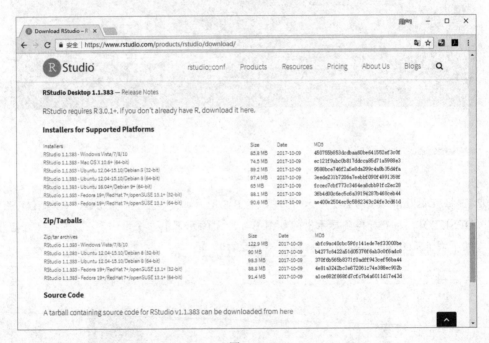

图 B-4

步骤 03 下载好之后启动安装程序,并按照如下步骤进行安装。

(1)单击安装程序文件,如图 B-5 所示。

图 B-5

(2)出现"用户账户控制"窗口,单击"是"按钮,就会出现如图 B-6 所示的"欢迎使用 'RStudio' 安装向导"窗口,单击"下一步"按钮以继续。

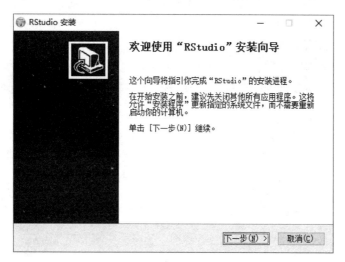

图 B-6

(3)单击"浏览"按钮,选择要安装的位置,接着单击"下一步"按钮,如图 B-7 所示。

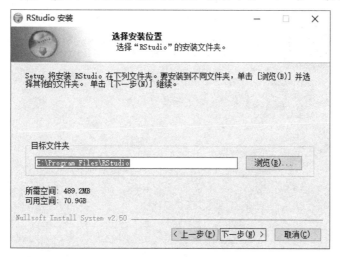

图 B-7

(4)出现"选择'开始菜单'文件夹"窗口,确定名称为"RStudio"后,单击"安装"按钮,如图 B-8 所示。

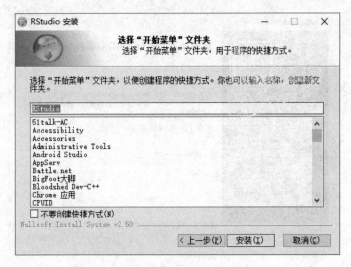

图 B-8

(5)接着就开始安装了,如图 B-9 所示。

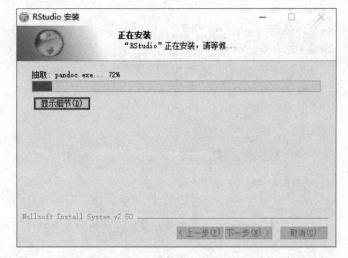

图 B-9

(6)完成安装后,就会显示如图 B-10 所示的界面。

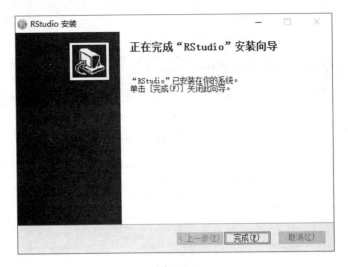

图 B-10

步骤 04 单击"完成"按钮之后,可以在"开始"→"RStudio"里找到安装好的 RStudio 软件,如图 B-11 所示。

图 B-11

步骤 05 启动 RStudio,会出现如图 B-12 所示的界面。

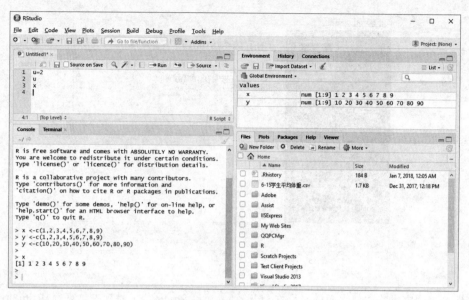

图 B-12

附录 C 安装 ODBC

Windows 操作系统

步骤 01 在 Windows 操作系统下设置 SQL Server（假设 IP 为 192.168.28.1），如图 C-1 所示。

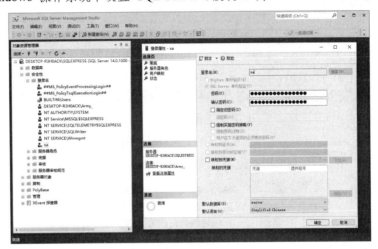

图 C-1

步骤 02 开启防火墙端口（port）1433 并在 SQL Server 配置管理器中启用 TCP/IP，如图 C-2 所示。

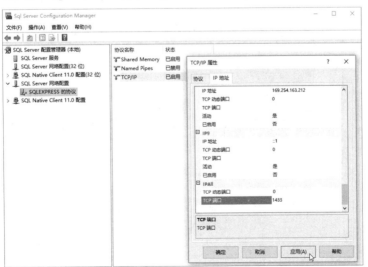

图 C-2

步骤03 在 Windows 10 的"控制面板"中启动"管理工具",再选择启动"设置 ODBC 数据源管理程序"(选择 64 位 或者 32 位)。具体设置步骤如图 C-3~图 C-8 所示。

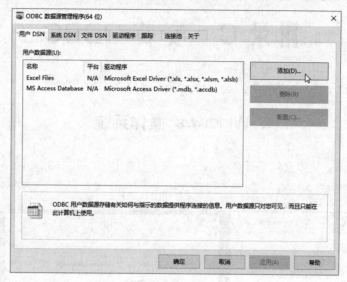

图 C-3

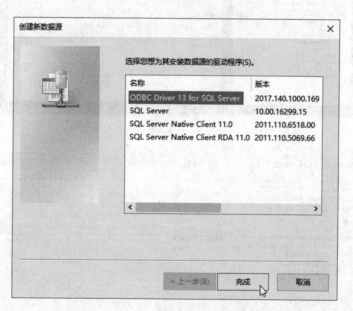

图 C-4

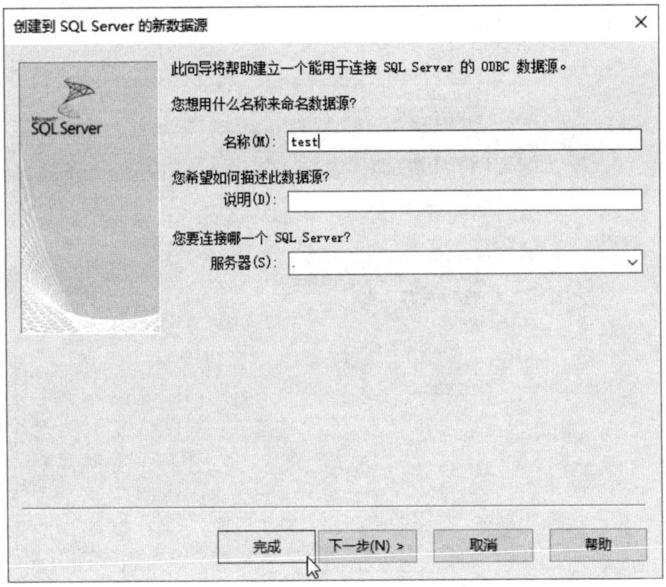
创建到 SQL Server 的新数据源
SQL Server
此向导将帮助建立一个能用于连接 SQL Server 的 ODBC 数据源。
您想用什么名称来命名数据源?
名称(M): test
您希望如何描述此数据源?
说明(D):
您要连接哪一个 SQL Server?
服务器(S): .
完成
下一步(N) >
取消
帮助

图 C-5

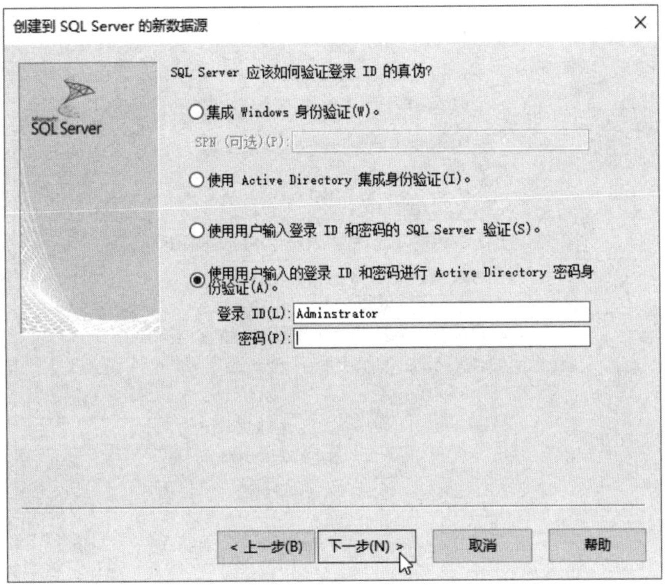
创建到 SQL Server 的新数据源
SQL Server
SQL Server 应该如何验证登录 ID 的真伪?
集成 Windows 身份验证(W)。
SPN (可选)(P):
使用 Active Directory 集成身份验证(I)。
使用用户输入登录 ID 和密码的 SQL Server 验证(S)。
使用用户输入的登录 ID 和密码进行 Active Directory 密码身份验证(A)。
登录 ID(L): Adminstrator
密码(P):
< 上一步(B)
下一步(N) >
取消
帮助

图 C-6

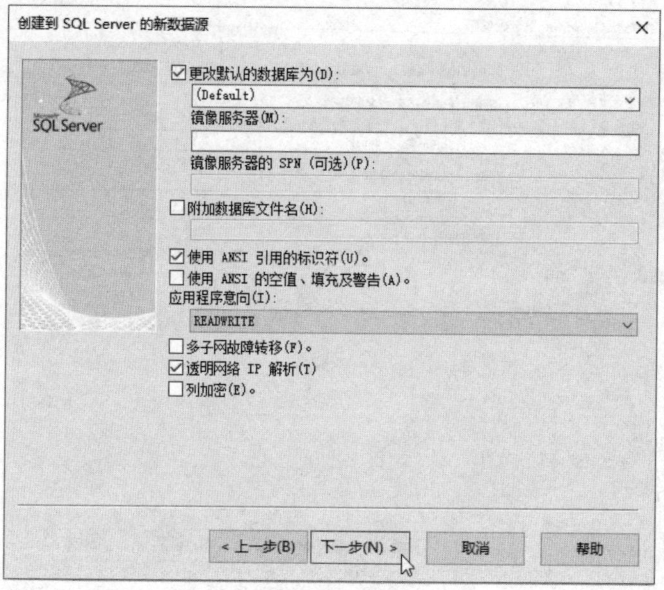

图 C-7

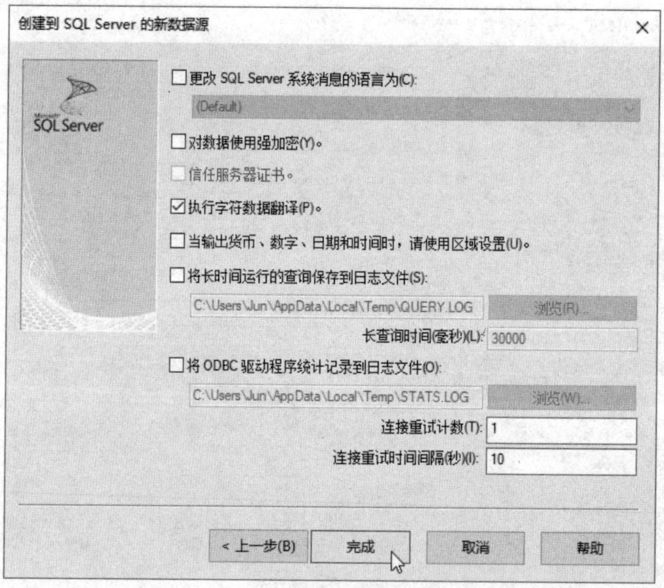

图 C-8

CentOS 操作系统

在 CentOS 操作系统中安装 ODBC：

```
# cd /root

# wget ftp://ftp.unixodbc.org/pub/unixODBC/unixODBC-2.3.0.tar.gz
```

```
# tar -zxvf unixODBC-2.3.0.tar.gz

# cd unixODBC-2.3.0/

# CPPFLAGS="-DSIZEOF_LONG_INT=8"
# export CPPFLAGS

# ./configure --prefix=/usr --libdir=/usr/lib64 --sysconfdir=/etc --enable- gui=no --enable-drivers=no --enable-iconv --with-iconv-char-enc=UTF8 --with- iconv-ucode-enc=UTF16LE

# make
# make install

# wget https://download.microsoft.com/download/B/C/D/BCDD264C-7517-4B7D-8159-C99FC5535680/RedHat6/msodbcsql-11.0.2270.0.tar.gz

# tar -zxvf msodbcsql-11.0.2270.0.tar.gz

# cd msodbcsql-11.0.2270.0

# ./install.sh install --lib-dir=/usr/local/lib64 --accept-license

# more /etc/odbcinst.ini
```

检查是否为下列信息：

```
[ODBC Driver 11 for SQL Server]
Description=Microsoft ODBC Driver 11 for SQL Server
Driver=/usr/local/lib64/libmsodbcsql-11.0.so.2270.0
Threading=1
UsageCount=1

# vi /etc/odbc.ini
[test]
Driver=ODBC Driver 11 for SQL Server #需要与 odbcinst.ini中的 driver_name一样
Description=My Sample ODBC Database Connection
Trace=Yes
Server=192.168.28.1
Port=1433
Database=test
```

测试 ODBC 连接：

```
# isql -v test test test

SQL > select * from iris
```

附录D 指令及用法

指令	功能
基本操作	
demo(程序包)	展示程序包的示范功能
library(程序包)	加载程序包
data(数据文件)	加载数据文件
head(数据文件,n)	显示数据文件前 n 项数据，默认 n=6
tail(数据文件,n)	显示数据文件后 n 项数据，默认 n=6
setwd("路径")	改变工作目录
getwd()	获取工作目录
length()	计算对象中元素的数量
mode()	获取对象的数据类型
class()	获取对象的类
str()	获取对象的数据结构
rm(list=ls());gc()	清除所有对象
对象及其运算	
%/%	整除
%%	求余数
%*%	矩阵相乘
%in%	判断是否在某个集合内
|	或（OR）
&	且（AND）
!	否（NOT）
seq(from, to, by)	产生以 by 为递增值的向量
assign("对象名称", 表达式)	创建对象，并将表达式值存入
c()	创建向量对象
array(x, dim=c())	按照 dim 维数创建 array 对象
matrix(x, nrow, ncol, byrow)	创建 x 为 nrow × ncol 矩阵，byrow=T 时以行的顺序排列

指令	功能
factor(x, levels=c())	按照 levels 排列创建 factor 对象
data.frame()	创建 data.frame 对象
list(name1=value1,…)	创建 list 对象
rbind()	按行合并
cbind()	按列合并
t()	转置矩阵
det()	求矩阵行列式的值
eigen()	计算矩阵特征值和特征向量
solve()	求逆矩阵
solve(A, b)	求矩阵 Ax=b 的 x 解
qr()	求矩阵 qr 分解
svd()	求矩阵 svd 分解
edit()	以电子表格方式编辑对象数据
view()	以电子表格方式显示对象数据
fix()	以电子表格方式修改对象数据
read.table()	输入多种格式的数据文件
scan()	数据输入
read.csv()	输入逗号分隔的 CSV 格式文件
write.table()	输出多种格式的数据文件
write.csv()	输出逗号分隔的 CSV 格式文件
save()	将对象保持为 Rdata 格式
load()	加载 Rdata 文件内的所有对象
file.choose()	以窗口选择来替代路径，搭配输入函数
as.	各种对象的转换
流程控制及循环	
ifelse(condition, T, F)	用于二分类逻辑判断，condition 成立时执行 T，否则执行 F
if(condition){程序语句 1}else{程序语句 2}	condition 成立时执行程序语句 1，否则执行程序语句 2
switch(计算值,程序语句 1, 程序语句 2…)	按照计算值（整数或文字）决定要执行的程序语句
while(condition){程序语句}	condition 成立时执行程序语句
repeat{}	重复执行程序语句直到跳出（break 语句跳出）

指令	功能
break	终止并跳离循环
next	跳过其后的运算，直接执行下一轮循环
apply(x,MARGIN,function)	对矩阵或数据框的 x 对象按行或列（MARGIN=1 或 2）执行 function 函数
lapply(x ,function)	对 x 对象执行 function 函数并以 list 方式返回
sapply(x, function)、	对 x 对象执行 function 函数并返回较简单的向量或矩阵
数学函数	
sum()	返回元素总和
abs(x)	返回 x 绝对值
sqrt(x)	返回 x 开根号值
ceiling(x)	返回大于等于 x 的最小整数
floor(x)	返回小于等于 x 的最大整数
round(x,n)	将 x 四舍五入到第 n 位
trunc(x)	返回 x 的整数部分
max()	返回最大值
min()	返回最小值
sign()	判断正负号
exp()	指数函数
log(x,base)	对数函数
sin();cos();tan();asin();acos;atan()	三角函数，反三角函数
mean()	返回平均值
median()	返回中位数
var()	返回方差
sd()	返回标准差
range()	返回最大及最小值
IQR()	返回 IQR 值
cor(x, y)	返回 x 及 y 的相关系数
绘图函数	
plot(x)	以序号为横坐标（x 轴）、y 为纵坐标（y 轴）来绘图
plot(x, y)	以 x（x-轴）和 y（y-轴）来绘图
pie(x)	绘制饼图
boxplot(x)	绘制盒形图

指令	功能
stem(x)	绘制茎叶图
dotchart(x)	绘制点图
hist(x)	绘制直方图
barplot(x)	绘制条形图
contour(x, y, z)	绘制等高线图
points(x, y)	画点
lines(x,y)	画直线
text(x, y, labels)	在指定位置写出指定文字
abline(a, b)	画出直线(y=ax+b)
abline(h=y);abline(v=x)	画出水平线或垂直线
legend(x,y, legend)	在指定位置画出图例
title(main,sub)	加上主标题或副标题
locator(n, type)	单击当前图形上的特定位置，最多单击 n 次
identify(x, y, n, label)	在指定点旁显示其在原向量中的指标值，最多点取 n 次
par(margin=c(a, b ,c, d))	设置离底部、左边、上方及右边的边界值，单位为英寸
par(mfcol=c(a, b))	以 a×b 矩阵将多张图形画在同一页，按列的顺序画图
par(mfrow=c(a, b))	以 a×b 矩阵将多张图形画在同一页，按行的顺序画图
其他函数	
na.omit();na.exclude()	删除 NA
is.na()	判断元素是否为 NA
na.rm=T	在某些函数内使用，可删除 NA
proc.time();system.time()	测量程序代码运行时间
system("command")	调用系统函数
Sys.time()	读取系统时间
Sys.sleep(x)	让程序暂时停止执行 x 秒

附录 E 在虚拟机上安装 R+Hadoop

本书提供了单机版的 R+Hadoop 供用户练习，R+Hadoop 建立在 VMware Workstation 的虚拟机上，其安装软件版本如下（用户若要使用新版本，则必须先确认兼容性）：

VMware Workstation	10.X	RStusdo server	0.98.1079
CentOS	7.0-1406	R	3.1.1
Hadoop	2.5.1	rmr2	3.2.0
Hbase	0.98.7	rhbase	1.2.1
Hive	1.1.0	rhdfs	1.0.8
Sqoop	1.4.5	RHive	2.0-0.2

安装 CentOS 和 Hadoop

创建虚拟机，步骤参考图 E-1~图 E-5。

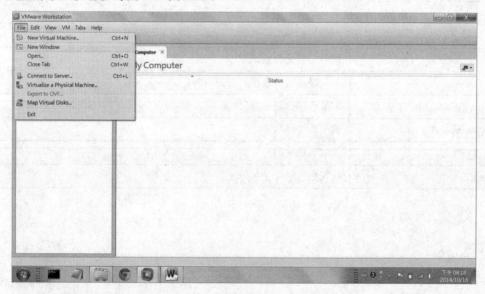

图 E-1

图 E-2

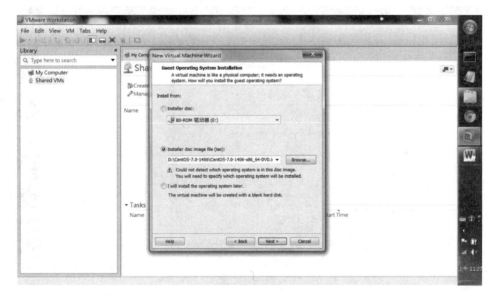

图 E-3

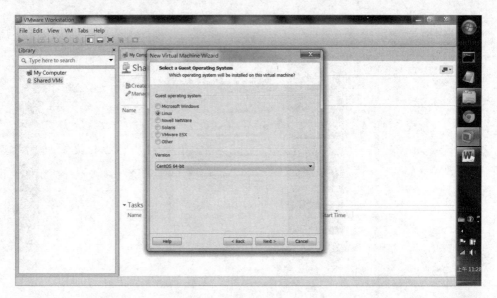

图 E-4

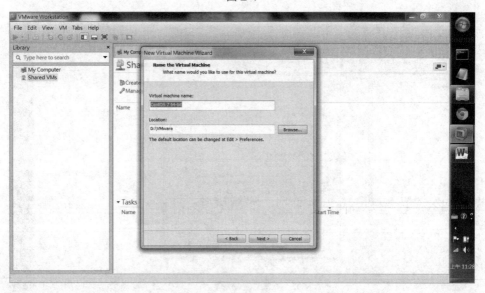

图 E-5

选择将 VMware Location 创建于 D:\VMware，如图 E-6 所示。

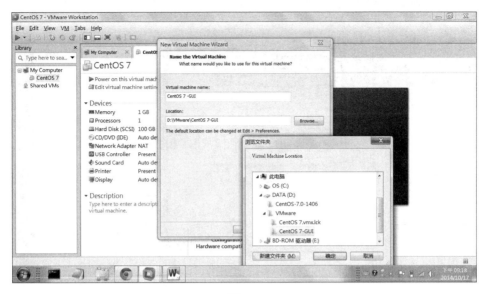

图 E-6

设置 disk size（硬盘空间）为 100 GB，如图 E-7 和图 E-8 所示。

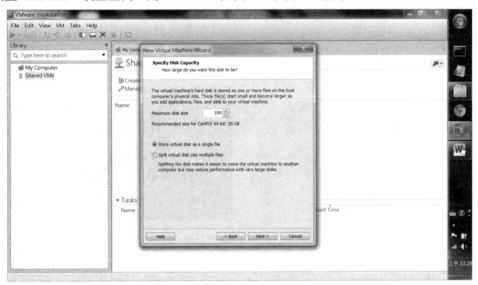

图 E-7

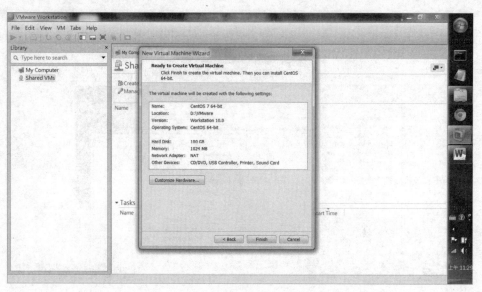

图 E-8

选择 Start Up Guest 启动访客模式，如图 E-9 和图 E-10 所示。

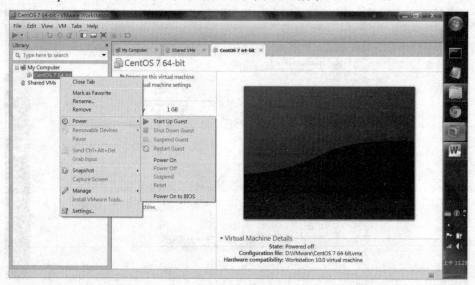

图 E-9

图 E-10

VMware 网络设置

使用 192.168.244.0 网段，如图 E-11 所示。

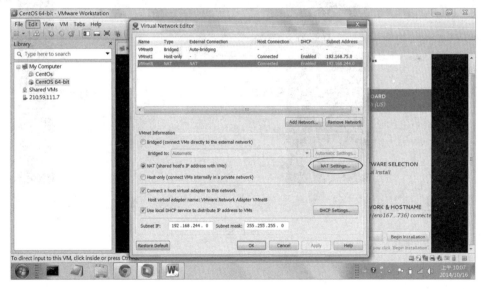

图 E-11

设置 DHCP 开始与结束 IP 地址，如图 E-12 所示。

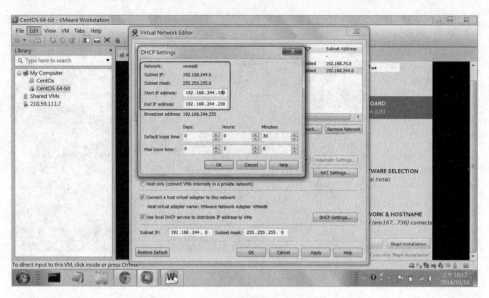

图 E-12

开始安装 CentOS 7 虚拟机并设置 DATE & TIME，如图 E-13 和图 E-14 所示。

图 E-13

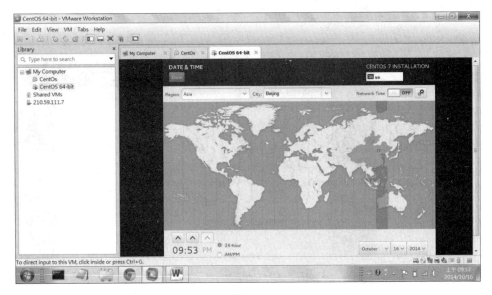

图 E-14

单击"INSTALLATION SOURCE"选项后再单击"Done"按钮,如图 E-15 和图 E-16 所示。

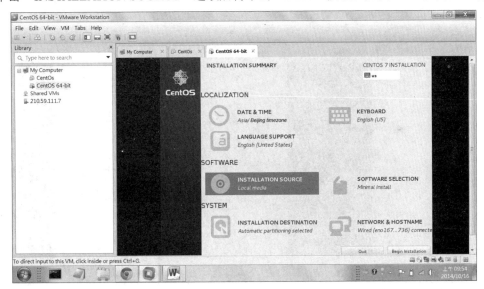

图 E-15

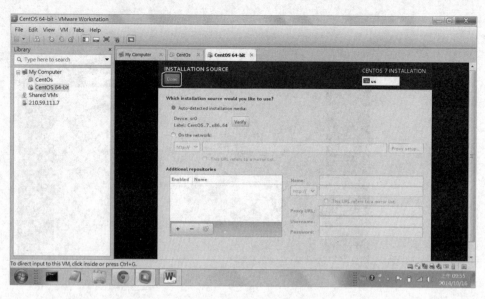

图 E-16

单击"SOFTWARE SELECTION"选项,如图 E-17 所示。

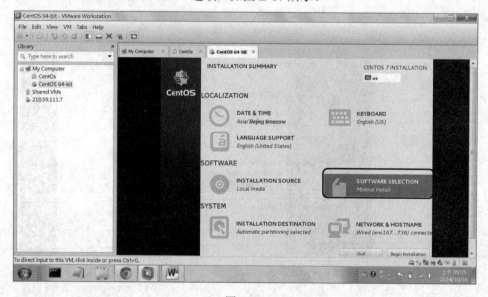

图 E-17

选择"GNOME Desktop"选项右边的各个安装项目,如图 E-18 所示。

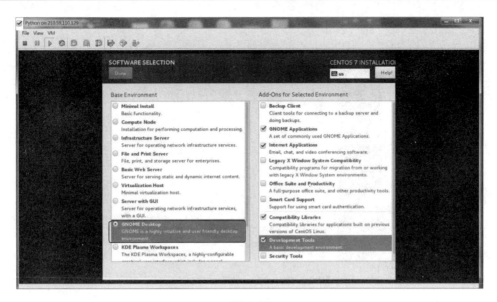

图 E-18

单击"INSTALLATION DESTINATION"选项,然后单击"Done"按钮,如图 E-19 和图 E-20 所示。

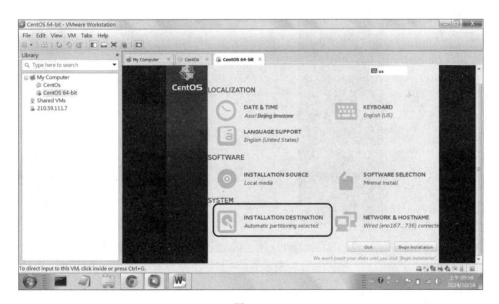

图 E-19

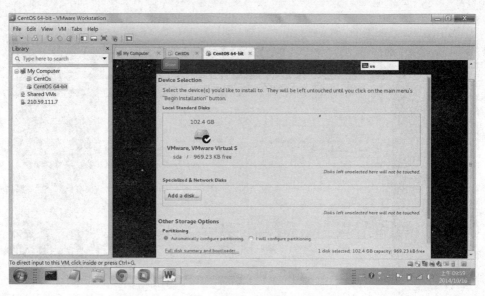

图 E-20

单击"NETWORK & HOSTNAME"选项来设置网络,如图 E-21 所示。

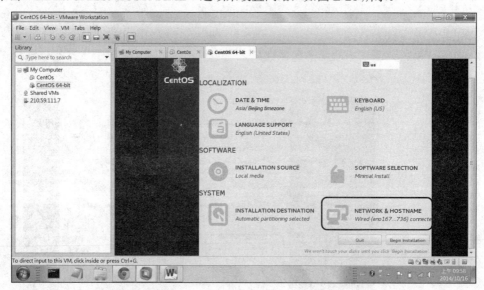

图 E-21

单击"ON"按钮以启动网络,如图 E-22 所示。

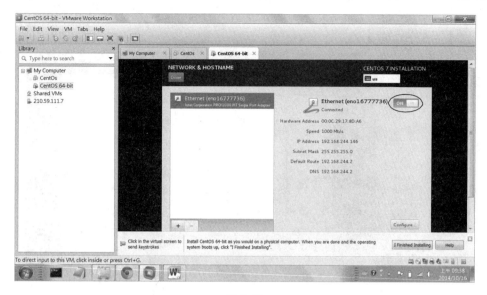

图 E-22

确认"General"中只勾选第二个选项，如图 E-23 所示。

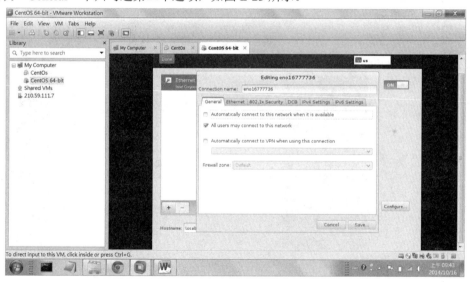

图 E-23

设置 IPv4（用户可自行设置 IP 地址），如图 E-24 所示。

Address 192.168.244.170

Netmask 255.255.255.0

Gateway 192.168.244.2

DNS 8.8.8.8

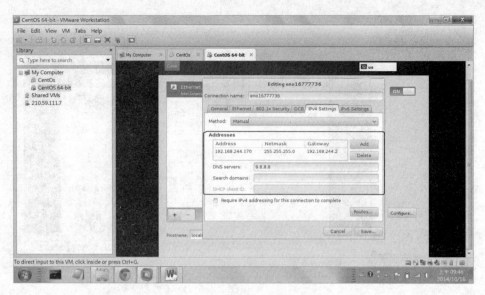

图 E-24

单击"Begin Installation"按钮,如图 E-25 所示。

图 E-25

单击"Root Password"选项,如图 E-26 所示。

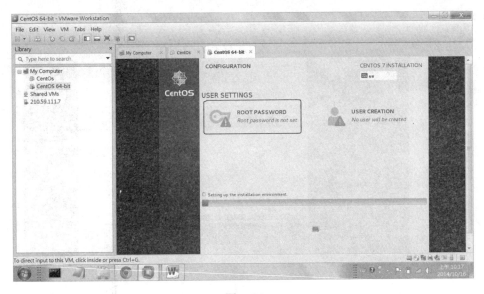

图 E-26

设置 Root Password 为 hadoop，如图 E-27 所示。

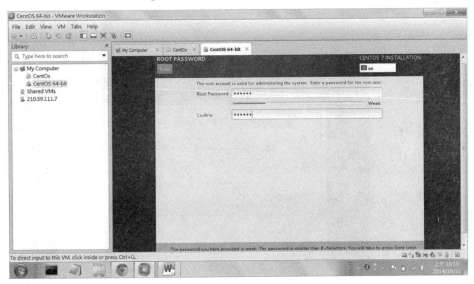

图 E-27

单击"USER CREATION"选项，如图 E-28 所示。

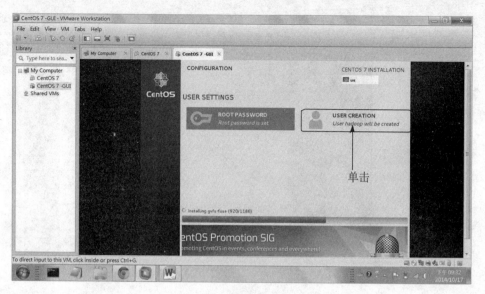

图 E-28

设置 USER CREATION 的 Username 为 hadoop、Password 为 hadoop，并勾选 Username 下的两个复选框，如图 E-29 所示。

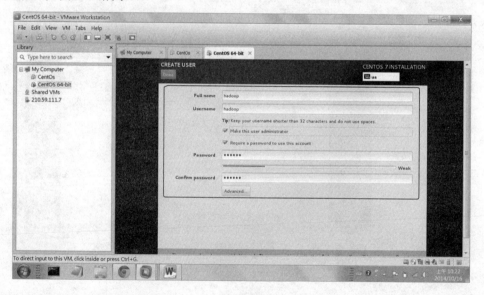

图 E-29

开始安装，如图 E-30 所示。

图 E-30

安装完毕后单击"Reboot"按钮,如图 E-31 所示。

图 E-31

安装 Hadoop

官方网站地址为 http://hadoop.apache.org/。

下载地址为 http://www.apache.org/dyn/closer.cgi/hadoop/common/。

以 hadoop/hadoop 账号和密码登录:

增加 hadoop可执行 sudo权限

```
visudo
#root    ALL=(ALL) ALL 在此行之下增加
hadoop   ALL=(ALL) ALL
```

先关闭 SELinux (Security Linux) 和 iptables, 再关闭 SELinux:

```
setenforce 0
```

设置 reboot 后自动关闭 SELinux。

用 Vi 编辑器打开配置文件:

```
vi /etc/selinux/config
```

找到:

```
SELINUX=
```

设置 SELINUX=disabled:

```
vi /etc/selinux/config
SELINUX=disabled
```

关闭 iptables:

```
service iptables stop
service ip6tables stop
```

关掉防火墙:

```
systemctl   stop   firewalld
```

设置 reboot 后自动关闭 iptable:

```
chkconfig iptables off
chkconfig ip6tables off

systemctl   disable   firewalld
systemctl   status    firewalld
```

安装 Java (版本会更新):

```
cd /usr/lib/jvm
yum -y install   java
yum -y install   java-1.7.0-openjdk-devel
```

确认 Java 安装后的版本:

```
java -version
ls          .
```

设置 JAVA_HOME:

```
echo 'export JAVA_HOME=/usr/lib/jvm/java-1.7.0-openjdk-1.7.0.71-2.5.3.1.el7_0. x86_64/jre' >> /etc/profile
```

P.S. java-1.7.0-openjdk-1.7.0.71-2.5.3.1.el7_0.x86_64 需与 ls 的 Java 文件相同:

source /etc/profile

安装 wget：

```
yum install wget
cd /usr/local
wget http://apache.stu.edu.tw/hadoop/common/hadoop-2.5.1/hadoop-2.5.1.tar.gz

tar zxvf hadoop-2.5.1.tar.gz
echo 'export HADOOP_HOME=/usr/local/hadoop-2.5.1' >> /etc/profile echo 'export PATH=$PATH:$HADOOP_HOME/bin' >> /etc/profile
echo 'export PATH=$PATH:$HADOOP_HOME/sbin' >> /etc/profile source /etc/profile

chown -R hadoop /usr/local/hadoop-2.5.1
```

查看当前情况：

```
[root@localhost local]# hadoop version
Hadoop 2.5.1
Subversion https://git-wip-us.apache.org/repos/asf/hadoop.git -r 2e18d179e4a8065b6a9f29cf2de9451891265cce
Compiled by jenkins on 2014-09-05T23:11Z
Compiled with protoc 2.5.0
From source with checksum 6424fcab95bfff8337780a181ad7c78
This command was run using /usr/local/hadoop-2.5.1/share/hadoop/common/hadoop-common-2.5.1.jar
[root@localhost jvm]#


[root@localhost local]# hadoop
Usage: hadoop [--config confdir] COMMAND
where COMMAND is one of:
  fs                   run a generic filesystem user client
  version              print the version
  jar <jar>            run a jar file
  checknative [-a|-h]  check native hadoop and compression libraries availability
  distcp <srcurl> <desturl> copy file or directories recursively
  archive -archiveName NAME -p <parent path> <src>* <dest> create a hadoop archive
  classpath            prints the class path needed to get the
                       Hadoop jar and the required libraries
  daemonlog            get/set the log level for each daemon or
  CLASSNAME            run the class named CLASSNAME

Most commands print help when invoked w/o parameters.
[root@localhost jvm]#


[root@localhost local]# hdfs dfs -ls
Found 11 items
drwxr-xr-x   - root root       4096 2014-10-17 11:07 java

drwxr-xr-x   - root root       4096 2014-10-17 11:07 java-1.7.0
```

```
drwxr-xr-x   - root root    4096 2014-10-17 11:07 java-1.7.0-openjdk
drwxr-xr-x   - root root    4096 2014-10-17 11:07 java-1.7.0-openjdk-
1.7.0.51-2.4.5.5.el7.x86_64
-rw-r--r--   1 root root    9574020    2014-10-17    11:02    java-1.7.0-openjdk-devel-
1.7.0.51-2.4.5.5.el7.x86_64.rpm
drwxr-xr-x   - root root    4096 2014-10-17 11:07 java-openjdk
drwxr-xr-x   - root root    26 2014-10-17 09:30 jre
drwxr-xr-x   - root root    26 2014-10-17 09:30 jre-1.7.0
drwxr-xr-x   - root root    26 2014-10-17 09:30 jre-1.7.0-openjdk
drwxr-xr-x   - root root    26    2014-10-17    09:30    jre-1.7.0-openjdk-
1.7.0.51-2.4.5.5.el7.x86_64
drwxr-xr-x   - root root    26 2014-10-17 09:30 jre-openjdk
```

安装 Apache Hbase

Hbase 是搭配在 Hadoop 上的 Database，主要是 implement Google 的 BigTable。
安装 Hbase（hbase-0.98.7-hadoop2）：

```
cd /usr/local
wget
http://archive.apache.org/dist/hbase/hbase-0.98.7//hbase/hbase-0.98.7/hbase- 0.98.7-hadoop2
-bin.tar.gz
tar -zxvf hbase-0.98.7-hadoop2-bin.tar.gz
```

在 /usr/local /hbase-0.98.7-hadoop2/conf/hbase-site.xml 中加入：

```
vi /usr/local /hbase-0.98.7-hadoop2/conf/hbase-site.xml
<configuration>
<property>
<name>hbase.rootdir</name>
<value>file:///home/hbase/local/hbase</value>
</property>
<property>
<name>hbase.zookeeper.property.dataDir</name>
<value>/home/local/var/zookeeper</value>
</property>
</configuration>
```

修改 /usr/local /hbase-0.98.7-hadoop2/conf/hbase-env.sh：

```
vi /usr/local/hbase-0.98.7-hadoop2/conf/hbase-env.sh
export JAVA_HOME=/usr/lib/jvm/java-1.7.0-openjdk-1.7.0.71-2.5.3.1.el7_0. x86_64/jre
```

设置与 Sqoop 相关的参数：

```
vi /usr/local/sqoop/conf/sqoop-env.sh
export HBASE_HOME=/usr/local/hbase-0.98.7-hadoop2
```

环境变量设置：

```
vi /etc/profile
export HBASE_HOME=/usr/local/hbase-0.98.7-hadoop2
export PATH=$HBASE_HOME/bin:$PATH
```

保存文件后退出 vi 程序：

```
source /etc/profile
```

测试 start-hbase.sh：

```
cd /usr/local start-hbase.sh
[root@localhost local]# cd /usr/local [root@localhost local]# start-hbase.sh master running as process 3959. Stop it first. [root@localhost local]#
```

测试 hbase 功能：

```
[root@localhost local]# hbase shell
2014-10-21 09:50:55,695 INFO [main] Configuration.deprecation: hadoop.native. lib is deprocated. Instead, use io.native.lib.available
HBase Shell; enter 'help<RETURN>' for list of supported commands. Type "exit<RETURN>" to leave the HBase Shell
Version 0.98.7-hadoop2, r800c23e2207aa3f9bddb7e9514d8340bcfb89277, Wed Oct 8 15:58:11 PDT 2014


hbase(main):001:0> help
HBase Shell, version 0.98.7-hadoop2, r800c23e2207aa3f9bddb7e9514d8340bcfb89277, Wed Oct 8 15:58:11 PDT 2014
Type 'help "COMMAND"', (e.g. 'help "get"' -- the quotes are necessary) for help on a specific command.
Commands are grouped. Type 'help "COMMAND_GROUP"', (e.g. 'help "general"') for help on a command group.


COMMAND GROUPS: Group name: general
Commands: status, table_help, version, whoami
Group name: ddl
Commands: alter, alter_async, alter_status, create, describe, disable, disable_all, drop, drop_all, enable, enable_all, exists, get_table, is_ disabled, is_enabled, list, show_filters


Group name: namespace
Commands: alter_namespace, create_namespace, describe_namespace, drop_ namespace, list_namespace, list_namespace_tables


Group name: dml
Commands: append, count, delete, deleteall, get, get_counter, incr, put, scan, truncate, truncate_preserve


Group name: tools
```

```
    Commands: assign, balance_switch, balancer, catalogjanitor_enabled, catalogjanitor_run,
catalogjanitor_switch, close_region, compact, flush, hlog_roll, major_compact, merge_region,
move, split, trace, unassign, zk_dump
    Group name: replication
    Commands: add_peer, disable_peer, enable_peer, list_peers, list_replicated_ tables,
remove_peer, set_peer_tableCFs, show_peer_tableCFs

    Group name: snapshots
    Commands: clone_snapshot, delete_snapshot, list_snapshots, rename_snapshot, restore_snapshot,
snapshot

    Group name: security
    Commands: grant, revoke, user_permission

    Group name: visibility labels
    Commands: add_labels, clear_auths, get_auths, set_auths, set_visibility

    SHELL USAGE:
    Quote all names in HBase Shell such as table and column names. Commas  delimit  command
parameters.   Type <RETURN> after entering a command to run it. Dictionaries of configuration used
in the creation and alteration of
    tables are
    Ruby Hashes. They look like this:

    {'key1' => 'value1', 'key2' => 'value2', ...}

    and are opened and closed with curley-braces. Key/values  are  delimited  by  the  '=>'
character combination. Usually keys are predefined constants such as NAME, VERSIONS, COMPRESSION,
etc. Constants do not need to be quoted.
    Type
    'Object.constants' to see a (messy) list of all constants in the environment.

    If you are using binary keys or values and need to enter them in the shell, use double-quote'd
hexadecimal representation. For example:

    hbase> get 't1', "key\x03\x3f\xcd" hbase> get 't1', "key\003\023\011"
    hbase> put 't1', "test\xef\xff", 'f1:', "\x01\x33\x40"

    The HBase shell is the (J)Ruby IRB with the above HBase-specific commands added. For more
on the HBase Shell, see http://hbase.apache.org/docs/current/ book.html hbase(main):002:0> quit
    [root@localhost local]#
```

安装 R

步骤 01 安装 EPEL（Extra Packages for Enterprise Linux）Repository on CentOS 7.x。

```
wget http://dl.fedoraproject.org/pub/epel/7/x86_64/e/epel-release-7-2.noarch.rpm

rpm -ivh epel-release-7-2.noarch.rpm
```

```
Verify EPEL Installation
yum repolist | grep "^epel\|repo id"

[root@localhost ~]# yum repolist | grep "^epel\|repo id"
repo id                 repo name                                          status
epel/x86_64             Extra Packages for Enterprise Linux 7 - x86_64     6,158

[root@localhost ~]#
```

步骤02 安装 thrift-0.9.1。可到 http://thrift.apache.org/download 手动下载安装文件，再使用 winSCP 上传到 /usr/local。

安装步骤如下：

（1）安装 boost-devel：

```
yum install boost-devel
```

（2）安装 openssl-devel：

```
yum install openssl-devel
yum install automake libtool flex bison pkhonfig gcc-c++
```

（3）解压缩：

```
tar zxvf thrift-0.9.1.tar.gz
```

（4）执行 ./configure（在 thrift-0.9.1 文件夹下），确认 Building C++ Library 为 yes：

```
cd /usr/local/thrift-0.9.1
./configure thrift 0.9.1
Building C++ Library ......... : yes
Building C (GLib) Library .... : no
Building Java Library ........ : no
Building C# Library .......... : no
Building Python Library ...... : yes   Building Ruby Library ........ : no
Building Haskell Library ..... : no
Building Perl Library ........ : no
Building PHP Library ......... : no
Building Erlang Library ...... : no
Building Go Library .......... : no
Building D Library ........... : no

C++ Library:
Build TZlibTransport ...... : yes
Build TNonblockingServer .. : no
Build TQTcpServer (Qt) .... : no

Python Library:
Using Python .............. : /usr/bin/python
```

> If something is missing that you think should be present, please skim the output of configure to find the missing component. Details are present in config.log.

（5）修改/usr/local/thrift-0.9.1/lib/cpp/thrift.pc，将 includedir=${prefix}/include/修改为 includedir=${prefix}/include/thrift：

```
vi /usr/local/thrift-0.9.1/lib/cpp/thrift.pc
includedir=${prefix}/include/thrift
```

（6）到/usr/local/thrift-0.9.1/test/cpp 目录下执行 cp *.o / .libs（/后有空格，libs 是隐藏文件，可使用 lsattr –aR 将其内容显示出来），回到 thrift-0.9.1 目录再执行 make：

```
cd /usr/local/thrift-0.9.1/test/cpp
cp *.o / .libs

cd /usr/local/thrift-0.9.1
make
```

（7）make install，若无错误，即安装成功：

```
make install
```

（8）设置 pkg_config_path：export PKG_CONFIG_PATH=/usr/local/lib/pkgconfig：

```
export PKG_CONFIG_PATH=/usr/local/lib/pkgconfig
```

（9）安装完毕后执行 pkg-config --cflags thrift，结果如下即可：

```
-I /usr/local/include/thrift
pkg-config --cflags thrift
```

（10）复制 libthrift-0.9.1.so 文件到/usr/lib：

```
cp -rp /usr/local/lib/libthrift-0.9.1.so /usr/lib
```

（11）载入 libthrift-0.9.1.so 文件：

```
ldconfig /usr/lib/libthrift-0.9.1.so
```

注意：若 thrift 因故未安装成功，最好 make clean 或重新解压缩后再执行 ./configure 步骤。

步骤 03 安装 R：

```
yum install R
```

步骤 04 执行 R 软件：

```
R
```

步骤 05 在 R 中安装下列程序包（Package）：

```
install.packages(c("rJava","Rcpp","RJSONIO","bitops","digest","functional",
"stringr","plyr","reshape2","dplyr","R.methodsS3","caTools","Hmisc","rjson","bit64"))
```

安装好程序包后输入 q()即可退出 R：

```
q()
```

步骤 06 到 https://github.com/RevolutionAnalytics/RHadoop/wiki/Downloads 手动下载 ravro、rhdfs、rmr2 和 rhbase，再使用 WinSCP（免费软件，可到 http://winscp.net/eng/download.php 下载）上传到 /usr/local：

设置 Hadoop 环境变量：

```
vi /etc/profile
export HADOOP_HOME=/usr/local/hadoop-2.5.1
export PATH=$PATH:$HADOOP_HOME/bin
export PATH=$PATH:$HADOOP_HOME/sbin
export HADOOP_MAPRED_HOME=$HADOOP_HOME
export HADOOP_COMMON_HOME=$HADOOP_HOME
export HADOOP_CMD=$HADOOP_HOME/bin/hadoop
export HADOOP_HDFS_HOME=$HADOOP_HOME
```

退出 vi 后：

```
source /etc/profile
```

增加环境变量：

```
cat >> /usr/local/hadoop-2.5.1/libexec/hadoop-config.sh << EOF
export JAVA_HOME=/usr/lib/jvm/java-1.7.0-openjdk-1.7.0.71-2.5.3.1.el7_0.x86_64/jre
EOF
```

创建本地暂存的文件夹：

```
mkdir -p $HADOOP_HOME/tmp
```

确认 HADOOP_STREAMING 变量：

```
[root@localhost ~]# find /usr -name hadoop-streaming*
/usr/local/hadoop-2.5.1/share/doc/hadoop/hadoop-streaming
/usr/local/hadoop-2.5.1/share/hadoop/tools/lib/hadoop-streaming-2.5.1.jar
/usr/local/hadoop-2.5.1/share/hadoop/tools/sources/hadoop-streaming-2.5.1-test-sources.jar
/usr/local/hadoop-2.5.1/share/hadoop/tools/sources/hadoop-streaming-2.5.1-sources.jar
[root@localhost ~]#
[root@localhost ~]# echo 'export HADOOP_STREAMING=/usr/local/hadoop-2.5.1/share/hadoop/tools/lib/hadoop-streaming-2.5.1.jar' >> /etc/profile
source /etc/profile
[root@localhost ~]# reboot
```

以 hadoop/hadoop 账号和密码登录：

```
vi ~/.bashrc
export JVM_ROOT=/usr/lib/jvm

export JAVA_HOME=$JVM_ROOT/java-1.7.0-openjdk-1.7.0.71-2.5.3.1.el7_0.x86_64/jre

export HADOOP_HOME=/usr/local/Hadoop-2.5.1
export HADOOP_MAPRED_HOME=$HADOOP_HOME
```

```
export HADOOP_COMMON_HOME=$HADOOP_HOME
export HADOOP_CMD=$HADOOP_HOME/bin/hadoop
export HADOOP_HDFS_HOME=$HADOOP_HOME
export YARN_HOME=$HADOOP_HOME
export HADOOP_CONF_DIR=$HADOOP_HOME/etc/hadoop
export YARN_CONF_DIR=$HADOOP_HOME/etc/hadoop
export PATH=$PATH:$HADOOP_HOME/bin:$HADOOP_HOME/sbin
```

离开 vi 后：

```
source ~/.bashrc
```

检查 Hadoop 变量：

```
[root@localhost hadoop]# env | grep
HADOOP HADOOP_CMD=/usr/local/hadoop-2.5.1/bin/hadoop HADOOP_HOME=/usr/local/hadoop-2.5.1
    HADOOP_HDFS_HOME=/usr/local/hadoop-2.5.1    HADOOP_COMMON_HOME=/usr/local/hadoop-2.5.1
HADOOP_CONF_DIR=/usr/local/hadoop-2.5.1/etc/hadoop
    HADOOP_STREAMING=/usr/local/hadoop-2.5.1/share/hadoop/tools/lib/hadoop-
streaming-2.5.1.jar
    HADOOP_MAPRED_HOME=/usr/local/hadoop-2.5.1
```

安装 R+Hadoop 相关程序包：

```
cd/usr/local
R CMD INSTALL ravro_1.0.4.tar.gz
R CMD INSTALL rhdfs_1.0.8.tar.gz
R CMD INSTALL rmr2_3.2.0.tar.gz
R CMD INSTALL plyrmr_0.4.0.tar.gz
R CMD INSTALL rhbase_1.2.1.tar.gz
```

步骤07 安装 RStudio Server。

先下载 RStudio Server 版：

```
wget http://download2.rstudio.org/rstudio-server-0.98.1079-x86_64.rpm rpm -ivh --nodeps rstudio-server-0.98.1079-x86_64.rpm

ln -s /usr/lib64/libssl.so.10 /usr/lib64/libssl.so.6
ln -s /usr/lib64/libcrypto.so.10 /usr/lib64/libcrypto.so.6
/etc/init.d/rstudio-server start
```

步骤08 使用 RStudio。

在 Windows 下以 Chrome 打开 http://192.168.244.131:8787，并以 hadoop user 登录，若无此 USER，则添加此 user。

添加 user：

```
useradd hadoop -g hadoop
```

设置 user 密码：

```
passwd hadoop
```

安装 Hive & RHive

安装 Hive。

步骤 01 yum install wget
步骤 02 wget http://ftp.mirror.tw/pub/apache/hive/hive-0.14.0/ apache-hive- 0.14.0-bin.tar.gz
步骤 03 tar zxvf apache-hive-0.14.0-bin.tar.gz
步骤 04 cp -Rp apache-hive-0.14.0-bin /usr/local/hive
步骤 05 cd /usr/local/hive/conf
步骤 06 cp -Rp hive-default.xml.template hive-site.xml
步骤 07 vi hive-site.xml

```
<configuration>
<property>
<name>hive.querylog.location</name>
<value>/usr/local/hive/logs</value>
</property>
<property>
<name>hive.aux.jars.path</name>
<value>file:///usr/local/hive/lib/hive-hbase-handler-0.14.0.jar,
file:///usr/local/hive/lib/hbase-common-0.98.7-hadoop2.jar,
file:///usr/local/hive/lib/zookeeper-3.4.5.jar
</value>
</property>
<property>
<name>javax.jdo.option.ConnectionURL</name>
<value>jdbc:derby:;databaseName=/usr/local/hive/metastore_db;create=true</value>
</property>
<property>
<name>hive.server2.thrift.port</name>
<value>10000</value>
</property>
<property>
<name>hive.server2.thrift.bind.host</name>
<value>master</value>
</property>

</configuration>
```

步骤 08 cp -Rp /usr/local/hbase-0.98.7-hadoop2/lib/hbase-common-0.98.7- hadoop2.jar /usr/local/hive/lib
步骤 09 nohup /usr/local/hive/bin/hive --hiveserver2&

安装 RHive：

```
# R
> install.packages(rJava)
> install.packages(Rserve)
> install.packages(RHive)
> library(Rserve)
```

```
> library(RHive)
> Sys.setenv(HIVE_HOME="/usr/local/hive")
> Sys.setenv(HAPOOP_HOME="/usr/local/hadoop-2.5.1")
> rhive.init(hiveHome="/usr/local/hive",hiveLib="/usr/local/hive/lib",
hadoopHome="/usr/local/hadoop-2.5.1")
> rhive.connect('master', hiveServer2=TRUE)
> rhive.query('show databases')
```

安装 Apache Sqoop

选择 Sqoop 1.4.5 版本（http://ftp.twaren.net/Unix/Web/apache/sqoop/1.4.5/）：

```
cd  /usr/local

wget http://ftp.twaren.net/Unix/Web/apache/sqoop/1.4.5/sqoop-1.4.5.bin__
hadoop-2.0.4-alpha.tar.gz
```

下载后解压：

```
tar -zxvf sqoop-1.4.5.bin hadoop-2.0.4-alpha.tar.gz ln -s sqoop-1.4.5.bin hadoop-2.0.4-alpha sqoop
```

环境变量设置：

```
vi /etc/profile

export  SQOOP_HOME=/usr/local/sqoop  export PATH=$SQOOP_HOME/bin:$PATH
```

改变权限：

```
chmod 777 /usr/local/sqoop/bin/*
```

设置 Sqoop 参数：

```
cp /usr/local/sqoop/conf/sqoop-env-template.sh /usr/local/sqoop/conf/sqoop-env.sh

vi /usr/local/sqoop/conf/sqoop-env.sh
export HADOOP_COMMON_HOME=/usr/local/hadoop-2.5.1
export HADOOP_MAPRED_HOME=/usr/local/hadoop-2.5.1
```

复制 lib：

```
cp /usr/local/sqoop/sqoop-1.4.5.jar /usr/local/hadoop/lib/
```

测试 Sqoop：

```
sqoop help

Available commands:
Codegen Generate code to interact with database records
```

```
create-hive-table Import a table definition into Hive
eval Evaluate a SQL statement and display the results
export Export an HDFS directory to a database table
help List available commands
import Import a table from a database to HDFS
import-all-tables Import tables from a database to HDFS
job Work with saved jobs
list-databases List available databases on a server
list-tables List available tables in a database
merge Merge results of incremental imports
metastore Run a standalone Sqoop metastore
version Display version information

See 'sqoop help COMMAND' for information on a specific command.
[root@localhost etc]#
```

测试 Sqoop：

```
sqoop version

[root@localhost lib]# sqoop version
Warning: /usr/local/sqoop/../hbase does not exist! HBase imports will fail.
Please set $HBASE_HOME to the root of your HBase installation.
Warning: /usr/local/sqoop/../hcatalog does not exist! HCatalog jobs will fail.
Please set $HCAT_HOME to the root of your HCatalog installation.
Warning: /usr/local/sqoop/../accumulo does not exist! Accumulo imports will fail.
Please set $ACCUMULO_HOME to the root of your Accumulo installation.
Warning: /usr/local/sqoop/../zookeeper does not exist! Accumulo imports will fail.
Please set $ZOOKEEPER_HOME to the root of your Zookeeper installation. 14/10/19 10:04:47 INFO sqoop.Sqoop: Running Sqoop version: 1.4.5
    Sqoop 1.4.5
    git commit id 5b34accaca7de251fc91161733f906af2eddbe83 Compiled by abe on Fri Aug 1 11:19:26 PDT 2014 [root@localhost lib]#
```

安装 Microsoft SQL Server JDBC Driver

步骤 01 下载 Microsoft SQL Server JDBC 驱动程序（http://www.microsoft.com/en-us/download/details.aspx?displaylang=en&id=11774），并把它复制到目录：the/usr/local/sqoop/lib 中。

例如：

```
$ curl -L 'http://download.microsoft.com/download/0/2/A/02AAE597-3865-456C-AE7F-613F99F850A8/sqljdbc_4.0.2206.100_enu.tar.gz' | tar xz
$ sudo cp sqljdbc4.jar /usr/local/sqoop/lib
```

步骤 02 设置 SQL Server（假设 SQL Server 安装于 192.168.28.1），如图 E-32 所示。

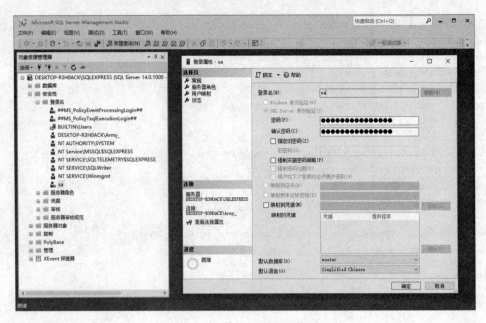

图 E-32

步骤 03 开启防火墙 port1433 并在 SQL Server 配置管理器中启用 TCP/IP，如图 E-33 所示。

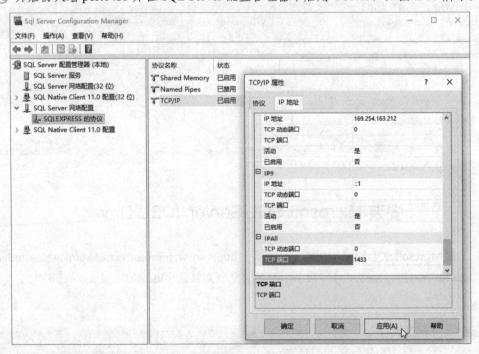

图 E-33

步骤 04 设置 Sqoop 指令：

```
sqoop import --connect "jdbc:sqlserver://192.168.28.1:1433;database=GDPSQL;username=sa;password=123456 --hive-import -m 1 --table GDP --warehouse-dir /user/hive/warehouse/mitopac --hive-overwrite
```

附录 F 在虚拟机上安装 SparkR

本书提供了 SparkR 安装步骤供读者参考，SparkR 创建在虚拟机上，其安装软件版本如下（用户若要使用新版本，则必须先确认兼容性）：

- CentOS   7.0-1406
- Spark   2.1
- Scala   2.11.8
- Cassandra 3.2
- Anaconda Python 2.7
- RStudio Server   1.0.136
- R   3.3.2

需注意以下内容：

（1）安装时下载源文件到 /home/spark/Downloads。
（2）应用程序安装到 /usr/local。

❖ 单机版

读者可先参考附录 D 创建 CentOS，再以 spark/spark 的账号/密码进入 CentOS，并切换到 root（使用 sudo 或 su -）：

```
$ su - Password: spark
```

❖ 增加 spark 的可执行 sudo 权限

```
# visudo

root ALL=(ALL)     ALL
spark    ALL=(ALL)     ALL
```

❖ 先检查当前的 hostname

```
# hostnamectl
```

❖ 将 hostname 修改为 master

```
# hostnamectl set-hostname master
# reboot
```

❖ 关闭 SELinux

```
# setenforce 0
```

❖ 设置 reboot 后自动关闭 SELinux

```
# vi /etc/selinux/config
SELINUX=disabled
```

❖ 关闭 iptables

```
# service iptables stop
# service ip6tables stop
```

❖ 关掉防火墙

```
# systemctl stop firewalld
```

❖ 设置 reboot 后自动关闭 iptable

```
# chkconfig iptables off
# chkconfig ip6tables off
# systemctl disable firewalld
# systemctl status firewalld
```

❖ 设置固定 IP 地址

以 ifconfig 找出网卡（Ex. eno16777736）：

```
# ifconfig

# vi /etc/sysconfig/network-scripts/ifcfg-eno16777736
DEVICE= eno16777736
BOOTPROTO=static
IPADDR=192.168.244.150 (使用者请自行设定IP网址)  IPV6INIT=no
IPV6_AUTOCONF=no
NETMASK=255.255.255.0
GATEWAY=192.168.244.2
ONBOOT=yes
DNS1=8.8.8.8
```

设置完成后，通过 service network restart 命令将整个网络服务重启，再查看重启后的网络配置（ifconfig）是否符合设置要求：

```
# service network restart
# ifconfig
```

❖ 使用程序包管理工具 yum 安装 Java（版本会更新）

```
# cd /usr/lib/jvm
# yum -y install java
# yum -y install java-1.8.0-openjdk-devel
```

❖ 确认 Java 安装后的版本

```
# java -version
# ls
```

❖ 设置 Java 路径

```
$ vi /etc/profile
export SCALA_HOME=/usr/local/scala
export PATH=$PATH:$SCALA_HOME/bin
export SPARK_HOME=/usr/local/spark export PATH=$PATH:$SPARK_HOME/bin
```

```
export JAVA_HOME=/usr/lib/jvm/java-1.8.0-openjdk-1.8.0.101-3.b13.el7_2.x86_64
export JAVA_ROOT=$JAVA_HOME/jre

export PATH="/root/anaconda2/bin:$PATH"
# source /etc/profile
# exit (离开 root)
$ vi ~/.bashrc
export SCALA_HOME=/usr/local/scala
export PATH=$PATH:$SCALA_HOME/bin
export SPARK_HOME=/usr/local/spark
export PATH=$PATH:$SPARK_HOME/bin

export JAVA_HOME=/usr/lib/jvm/java-1.8.0-openjdk-1.8.0.101-3.b13.el7_2.x86_64
export JAVA_ROOT=$JAVA_HOME/jre

export PATH="/root/anaconda2/bin:$PATH"
$ source ~/.bashrc
```

安装 Spark

先确认 Spark 的 Scala 版本：

```
# cd /home/spark/Downloads

# wget http://www.scala-lang.org/files/archive/scala-2.11.8.tgz

# wget http://d3kbcqa49mib13.cloudfront.net/spark-2.1.0-bin-hadoop2.7.tgz

# tar zxvf scala-2.11.8.tgz ; tar zxvf spark-2.1.0-bin-hadoop2.7.tgz

# cp -Rp scala-2.11.8 /usr/local
# cp -Rp spark-2.1.0-bin-hadoop2.7 /usr/local
```

若有需要，则先删除已安装的 Scala 和 Spark：

```
# cd /usr/local

# rm -rf scala
# rm -rf spark

# ln -s scala-2.11.8 scala
# ln -s spark-2.1.0-bin-hadoop2.7 spark
```

❖ 启动 Spark

```
$ /usr/local/spark/bin/spark-shell (scala)
$ /usr/local/spark/bin/pyspark --master local[2] (pyspark)
```

❖ SparkUI

http://192.168.244.150:4040（需执行 spark-shell 或 pyspark）

❖ 离开 Spark

```
scala> ctrl+c scala
>>> exit()
```

❖ 测试

```
# cd /usr/local/spark

# ./bin/spark-submit examples/src/main/python/pi.py 10
# ./bin/spark-submit examples/src/main/r/dataframe.R
```

❖ 增加执行 Spark 的权限

```
# chmod 755 /root

# cd /usr/local

# chown -R spark.spark spark
# chown -R spark.spark spark-2.1.0-bin-hadoop2.7

# chown -R spark.spark scala-2.11.8
# chown -R spark.spark scala
```

❖ 寻找 metastore_db 并执行 chmod

```
# find / -name metastore_db

# chmod -R a+rwx /root/metastore_db  (视实际情况）
# chmod -R a+rwx /usr/lib/jvm/metastore_db
# chmod -R a+rwx /home/spark/metastore_db
# chmod -R a+rwx /usr/local/spark-2.1.0-bin-hadoop2.7/metastore_db
```

安装 Jupyter-Notebook

这部分为安装 Python，读者可自行决定是否要安装 Python。

❖ 下载 Anaconda Python 2.7（网址为 https://www.anaconda.com/download/#_unix）

```
# cd /user/local
# wget http://repo.continuum.io/archive/Anaconda2-4.1.1-Linux-x86_64.sh
# bash Anaconda2-4.1.1-Linux-x86_64.sh
# vi /etc/profile
export PATH="/root/anaconda2/bin:$PATH"

# source /etc/profile
# jupyter notebook --generate-config
# cd
# cd .jupyter
```

```
# vi jupyter_notebook_config.py
c.NotebookApp.ip = '192.168.244.150'
c.NotebookApp.open_browser = False
c.NotebookApp.port = 8888
```

❖ 安装相关程序包

```
# conda install matplotlib
# conda install pandas

# yum install -y python-pip
# pip install graphviz    (不能使用 conda)
#   PYSPARK_DRIVER_PYTHON=ipython   PYSPARK_DRIVER_PYTHON_OPTS="notebook   --no-   browser"
/usr/local/spark/bin/pyspark --master local[2]
```

❖ 启动浏览器

```
http://192.168.244.150:8888
```

❖ 退出 pyspark

```
$ ctrl +C
```

安装 R

步骤 01 安装 EPEL（Extra Packages for Enterprise Linux）Repository on CentOS 7.x：

```
# wget http://dl.fedoraproject.org/pub/epel/7/x86_64/e/epel-release-7-5.noarch.rpm

# rpm -ivh epel-release-7-5.noarch.rpm Verify EPEL Installation
# yum repolist | grep "^epel\|repo id"

[root@localhost ~]# yum repolist | grep "^epel\|repo id"
repo id    repo name
status epel/x86_64 Extra Packages for Enterprise Linux 7 - x86_64

6,158
[root@localhost ~]#
```

步骤 02 安装 thrift-0.9.3。可到 http://thrift.apache.org/download 手动下载安装程序，再使用 winSCP 上传到 /usr/local。

❖ 安装步骤

（1）安装 boost-devel：

```
# yum install boost-devel
```

（2）安装 openssl-devel：

```
# yum install openssl-devel
# yum install automake libtool flex bison pkhonfig gcc-c++
```

（3）解压缩：

```
# tar zxvf thrift-0.9.3.tar.gz
```

（4）执行 ./configure（在 thrift-0.9.3 文件夹下），确认 Building C++ Library 为 yes：

```
# cd /usr/local/thrift-0.9.3
# ./configure thrift 0.9.3
Building C++ Library ......... :
yes Building C (GLib) Library .... : no
Building Java Library ........ : no
Building C# Library .......... : no
Building Python Library ...... : yes Building Ruby Library ........ : no
Building Haskell Library ..... : no
Building Perl Library ........ : no
Building PHP Library ......... : no
Building Erlang Library ...... : no
Building Go Library .......... : no
Building D Library ........... : no

C++ Library:
Build TZlibTransport ...... : yes
Build TNonblockingServer .. : no
Build TQTcpServer (Qt) .... : no

Python Library:
Using Python .............. : /usr/bin/python
```

If something is missing that you think should be present, please skim the output of configure to find the missing component. Details are present in config.log.

（5）修改 /usr/local/thrift-0.9.3/lib/cpp/thrift.pc，将 includedir=${prefix}/include/ 修改为 includedir=${prefix}/include/thrift：

```
# vi /usr/local/thrift-0.9.3/lib/cpp/thrift.pc
includedir=${prefix}/include/thrift
```

（6）到 thrift-0.9.3 目录中执行 make：

```
# cd /usr/local/thrift-0.9.3
# make
```

（7）make install，若无错误，即安装成功：

```
# make install
```

（8）设置 pkg_config_path：

```
export PKG_CONFIG_PATH=/usr/local/lib/pkgconfig
# export PKG_CONFIG_PATH=/usr/local/lib/pkgconfig
```

（9）安装完毕后执行 pkg-config --cflags thrift：

```
-I /usr/local/include/thrift
pkg-config --cflags thrift
```

（10）复制 libthrift-0.9.3.so 文件到/usr/lib：

```
cp -Rp /usr/local/lib/libthrift-0.9.3.so /usr/lib
```

（11）载入 libthrift-0.9.3.so 文件：

```
ldconfig /usr/lib/libthrift-0.9.3.so
```

注意，若 thrift 因故未安装成功，则最好 make clean 或重新解压缩后再执行 ./configure 步骤。

步骤 03 安装 R：

```
yum install R
```

步骤 04 执行 R 软件：

```
R
```

步骤 05 输入 q()退出 R：

```
q()
```

步骤 06 安装 SparkR 2.1 程序包：

```
# R CMD INSTALL /usr/local/spark/R/lib/SparkR -l /usr/lib64/R/library/
```

步骤 07 安装 R-STUDIO。

下载 R-STUDIO Server 版：

```
# wget https://download2.rstudio.org/rstudio-server-rhel-1.0.136-x86_64.rpm

# yum install --nogpgcheck rstudio-server-rhel-1.0.136-x86_64.rpm
```

重新启动 R-STUDIO：

```
# service rstudio-server stop
# service rstudio-server start（默认已启动）
```

步骤 08 使用 R-STUDIO。

在 Windows 下以 chrome 开启 http://192.168.244.150:8787。

安装 Cassandra

Cassandra 为 NoSQL 的一种，用户可自行决定是否要安装 Cassandra：

```
$ su -
# vi /etc/yum.repos.d/datastax.repo
[datastax-ddc]
name = DataStax Repo for Apache Cassandra
baseurl = http://rpm.datastax.com/datastax-ddc/3.2
```

```
enabled = 1
gpgcheck = 0


# yum install datastax-ddc


# vi /etc/cassandra/conf/cassandra.yaml cluster_name: 'Single Cluster'



num_tokens: 256
seed_provider:
# Addresses of hosts that are deemed contact points.
# Cassandra nodes use this list of hosts to find each other and learn
# the topology of the ring.  You must change this if you are running
# multiple nodes!
- class_name: org.apache.cassandra.locator.SimpleSeedProvider
parameters:
# seeds is actually a comma-delimited list of addresses.
# Ex: "<ip1>,<ip2>,<ip3>"
- seeds: "192.168.244.150"

listen_address: 192.168.244.150
rpc_address: 192.168.244.150
#endpoint_snitch: GossipingPropertyFileSnitch
endpoint_snitch: SimpleSnitch

# Whether to start the thrift rpc server.
start_rpc: true


# vi /etc/cassandra/conf/cassandra-rackdc.properties
dc=dc1
rack=rack1


# Add a suffix to a datacenter name. Used by the Ec2Snitch and Ec2MultiRegionSnitch
# to append a string to the EC2 region name.
#dc_suffix=


# Uncomment the following line to make this snitch prefer the internal ip when possible, as the Ec2MultiRegionSnitch does.
prefer_local=true


# vi /etc/profile


export CQLSH_NO_BUNDLED=true

clear Cassandra data(若需要调试)
# rm -rf /var/lib/cassandra/data/system/*
```

安装 Apache-Cassandra 与 R 的驱动程序

下载安装文件的方法如下:

```
# cd /usr/local
# wget http://www-eu.apache.org/dist/cassandra/3.7/apache-cassandra-3.7-bin.tar.gz
```

或

```
# wget https://archive.apache.org/dist/cassandra/3.7/apache-cassandra-3.7-bin.tar.gz

# tar zxvf apache-cassandra-3.7-bin.tar.gz

# cd /usr/local/apache-cassandra-3.7/lib
# wget https://storage.googleapis.com/google-code-archive-downloads/v2/apache-extras.org/cassandra-jdbc/cassandra-jdbc-1.2.5.jar

# wget http://repo1.maven.org/maven2/org/apache/cassandra/cassandra-all/1.1.0/cassandra-all-1.1.0.jar

# cd /usr/local
# chown -R spark.spark apache-cassandra-3.7
```

安装 Apache-Cassandra 与 Spark 的驱动程序（可先到 https://spark-packages.org/package/datastax/spark-cassandra-connector 检查合适版本）。

```
# cd /usr/local

# wget http://dl.bintray.com/spark-packages/maven/datastax/spark-cassandra-connector/2.0.0-M2-s_2.11/spark-cassandra-connector-2.0.0-M2-s_2.11.jar
```

若无目录，则先创建 /home/spark/.m2/repository/com/datastax/spark/spark-cassandra-connector/2.0.0-M2-s_2.11/ 目录，再复制 spark-cassandra-connector-2.0.0-M2-s_2.11.jar。

```
# cp spark-cassandra-connector-2.0.0-M2-s_2.11.jar /home/spark/.m2/repository/com/datastax/spark/spark-cassandra-connector/2.0.0-M2-s_2.11/

# cp spark-cassandra-connector-2.0.0-M2-s_2.11.jar /usr/local/spark/spark-cassandra-connector-2.0.0-M2-s_2.11.jar
```

❖ 启动 Cassandra

```
# su cassandra   /usr/sbin/cassandra
```

❖ 检查 cassandra node status

```
# nodetool status
```

❖ 启动 csql 并离开

```
# cqlsh 192.168.244.150 9042
exit
```

❖ 需要 Debug CQL 时

cqlsh 版本不对，检查路径：

```
# which cqlsh
/root/anaconda2/bin/cqlsh

# ls -l /usr/bin/cqlsh
-rwxr-xr-x. 1 root root 834 Jan 20 08:30 /usr/bin/cqlsh

以 /usr/bin/cqlsh取代 /root/anaconda2/bin/cqlsh
# mv /root/anaconda2/bin/cqlsh  /root/anaconda2/bin/cqlsh.o
# ln -s /usr/bin/cqlsh /root/anaconda2/bin/cqlsh

No module named cqlshlib
# find / -name cqlshlib
/usr/lib/python2.7/site-packages/cqlshlib

# cp /usr/lib/python2.7/site-packages/cqlshlib /usr/bin/
```

❖ 安装 Python-Cassandra Driver

```
# pip install cassandra-driver

# cqlsh 192.168.244.150 9042
```

❖ 使用 cassandra CQL 指令进行数据的导出/导入

（1）列出有哪些 keyspace：

```
DESCRIBE keyspaces;    // Output the names of all keyspaces.
```

（2）使用 keyspace：

```
CREATE KEYSPACE mykeyspace
WITH REPLICATION = { 'class' : 'SimpleStrategy', 'replication_factor' : 1 };
USE mykeyspace;
```

（3）创建 TABLE：

```
CREATE TABLE  Student (
NO   VARCHAR PRIMARY KEY,
NAME TEXT,
CLASS  VARCHAR
);
```

（4）输入数据：

```
INSERT INTO Student (no,name,CLASS) VALUES('B0001006', 'Frodo', 'IM1A');
```

（5）批次输入指令，类似 SQL STOREPROCEDURE：

```
BEGIN BATCH
INSERT INTO Student (no,name,class) VALUES('B0001007', 'Brodo', 'IM2A');
INSERT INTO Student (no,name,class) VALUES('B0001008', 'Krodo', 'IM3A');
```

```
INSERT INTO Student (no,name,class) VALUES('C0001018', '林大力 ', 'IM3A') ;
-->不能输入中文 APPLY BATCH;
```

（6）导出数据到 CSV 文件：

```
COPY Student (no,class,name)  TO 'temp_Student.csv';
```

（7）清空数据：

```
TRUNCATE Student;
```

（8）导入数据：

```
COPY Student (no,name,class) FROM 'temp_Student.csv';
```

（9）测试是否正常：

```
USE mykeyspace;

Select * from Student;
```

❖ 测试 Python-Cassandra(jupyter)

```
from cassandra.cluster import
Cluster cluster = Cluster(['192.168.244.150'])
session = cluster.connect()
session.set_keyspace('mykeyspace')
rows = session.execute('SELECT  * FROM Student;')
for user_row in rows:
print user_row.no, user_row.name;
```

启动方式

❖ 启动 PYSPARK

```
#PYSPARK_DRIVER_PYTHON=ipython  PYSPARK_DRIVER_PYTHON_OPTS="notebook  --no- browser"
/usr/local/spark/bin/pyspark --master local[2]
```

❖ 启动 Cassandra

```
# su cassandra   /usr/sbin/cassandra

# nodetool enablethrift
```

❖ 测试 Web 界面

```
http://192.168.244.150:8888   (PYSPARK)
http://192.168.244.150:4040   (Spark)
http://192.168.244.150:8787   (R)
```

Standalone 版

用户需先安装 CentOS 并按下述方式安装 master、slave-1 及 slave-2。

Node	Hostname	IP
1	master	192.168.244.151
2	slave-1	192.168.244.152
3	Slave-2	192.168.244.153

❖ 以 spark/spark 进入 CentOS

切换到 root（使用 sudo 或 su -）：

```
$ su -
Password: spark
```

增加 spark 可执行 sudo 的权限：

```
# visudo

root ALL=(ALL)    ALL
spark   ALL=(ALL)    ALL
```

先检查当前的 hostname：

```
# hostnamectl
```

将 192.168.244.151 的 hostname 修改为 master：

```
# hostnamectl set-hostname master
```

将 192.168.244.152 的 hostname 修改为 slave-1：

```
hostnamectl set-hostname slave-1
```

将 192.168.244.153 的 hostname 修改为 slave-2：

```
hostnamectl set-hostname slave-2
```

```
# reboot
```

关闭 SELinux：

```
# setenforce 0
```

设置 reboot 后自动关闭 SELinux：

```
# vi /etc/selinux/config SELINUX=disabled
```

关闭 iptables：

```
# service iptables stop
# service ip6tables stop
```

关掉防火墙:

```
# systemctl stop firewalld
```

设置 reboot 后自动关闭 iptable:

```
# chkconfig iptables off
# chkconfig ip6tables off
# systemctl disable firewalld
# systemctl status firewalld
```

❖ 设置固定 IP Address

以 ifconfig 找出网卡（Ex. ens160）:

```
# ifconfig
```

设置 master IP Address:

```
# vi /etc/sysconfig/network-scripts/ifcfg-ens160

DEVICE=ens160
BOOTPROTO=static
IPADDR=192.168.244.151
IPV6INIT=no
IPV6_AUTOCONF=no
NETMASK=255.255.255.0
GATEWAY=192.168.244.2
ONBOOT=yes
DNS1=8.8.8.8
```

设置 slave-1 IP Address:

```
# vi /etc/sysconfig/network-scripts/ifcfg-ens160
DEVICE=ens160
BOOTPROTO=static
IPADDR=192.168.244.152
IPV6INIT=no
IPV6_AUTOCONF=no NETMASK=255.255.255.0
GATEWAY=192.168.244.2
ONBOOT=yes
DNS1=8.8.8.8
```

设置 slave-2 IP Address:

```
# vi /etc/sysconfig/network-scripts/ifcfg-ens160

DEVICE=ens160
BOOTPROTO=static
IPADDR=192.168.244.153
IPV6INIT=no
IPV6_AUTOCONF=no NETMASK=255.255.255.0
GATEWAY=192.168.244.2
ONBOOT=yes
```

```
DNS1=8.8.8.8
```

设置完成后，通过 service network restart 命令将整个网络服务重启，再查看重启后的网络配置（ifconfig）是否符合设置要求：

```
# service network restart
# ifconfig
```

使用程序包管理工具 yum 安装 Java（版本会更新）：

```
# cd /usr/lib/jvm
# yum -y install java
# yum -y install java-1.8.0-openjdk-devel
```

确认 Java 安装后的版本：

```
# java -version
# ls
```

设置 Java 路径：

```
$ vi /etc/profile
export SCALA_HOME=/usr/local/scala
export PATH=$PATH:$SCALA_HOME/bin
export SPARK_HOME=/usr/local/spark
export PATH=$PATH:$SPARK_HOME/bin

export JAVA_HOME=/usr/lib/jvm/java-1.8.0-openjdk-1.8.0.91-0.b14.el7_2.x86_64
export JAVA_ROOT=$JAVA_HOME/jre

export PATH="/root/anaconda2/bin:$PATH"

# source /etc/profile
# exit (退出 root)

$ vi ~/.bashrc

export SCALA_HOME=/usr/local/scala
export PATH=$PATH:$SCALA_HOME/bin
export SPARK_HOME=/usr/local/spark
export PATH=$PATH:$SPARK_HOME/bin

export JAVA_HOME=/usr/lib/jvm/java-1.8.0-openjdk-1.8.0.91-0.b14.el7_2.x86_64
export JAVA_ROOT=$JAVA_HOME/jre

export PATH="/root/anaconda2/bin:$PATH"

$ source ~/.bashrc
```

设置 ssh config：

```
$who spark
```

```
$ ssh-keygen -t rsa -f ~/.ssh/id_rsa -P ""
$ cp ~/.ssh/id_rsa.pub ~/.ssh/authorized_keys
$ scp -r ~/.ssh slave-1:~/
$ scp -r ~/.ssh slave-2:~/
```

测试从 master ssh to slaves：

```
$ ssh slave-1 (直接登录即成功)

#who
root
# ssh-keygen -t rsa -f ~/.ssh/id_rsa -P ""
# cp ~/.ssh/id_rsa.pub ~/.ssh/authorized_keys
# scp -r ~/.ssh slave-1:~/
# scp -r ~/.ssh slave-2:~/
```

测试从 master ssh to slaves：

```
# ssh slave-1 (直接登录即成功)
```

安装 Spark

先确认 Spark 的 Scala 版本：

```
# cd /home/spark/Downloads
# wget http://www.scala-lang.org/files/archive/scala-2.11.8.tgz

# wget http://d3kbcqa49mib13.cloudfront.net/spark-2.1.0-bin-hadoop2.7.tgz

# tar zxvf scala-2.11.8.tgz
# tar zxvf spark-2.1.0-bin-hadoop2.7.tgz

# cp -Rp scala-2.11.8 /usr/local
# cp -Rp spark-2.1.0-bin-hadoop2.7 /usr/local

# cd /usr/local
```

若有需要，则移除旧版本 Scala 及 Spark：

```
# rm -rf scala
# rm -rf spark

# ln -s scala-2.11.8 scala
# ln -s spark-2.1.0-bin-hadoop2.7 spark
```

启动 Spark：

```
$ /usr/local/spark/bin/spark-shell (scala)
$ /usr/local/spark/bin/pyspark --master local[2](pyspark)
```

SparkUI：

```
http://192.168.244.151:4040 (需执行 spark-shell 或 pyspark) http://192.168.244.152:4040
(需执行 spark-shell 或 pyspark) http://192.168.244.153:4040 (需执行 spark-shell 或
pyspark)
```

退出 Spark：

```
scala> ctrl+c    scala
>>> exit()      python
```

测试：

```
# cd /usr/local/spark

# ./bin/spark-submit examples/src/main/python/pi.py 10
# ./bin/spark-submit examples/src/main/r/dataframe.R
```

安装 Standalone Spark

步骤 01 确认 /root 权限（permission）为 755：

```
# chmod 755 /root
```

步骤 02 添加 master 和 slaves 的域名解析对应关系：

```
# vi /etc/hosts
192.168.244.151 master
192.168.244.152 slave-1
192.168.244.153 slave-2
```

步骤 03 增加执行 spark 及 mesos 的权限：

```
# cd /usr/local

# chown -R spark.spark spark
# chown -R spark.spark spark-2.1.0-bin-hadoop2.7

# chown -R spark.spark scala-2.11.8
# chown -R spark.spark scala

# chown -R spark.spark mesos
# chown -R spark.spark mesos-0.27.2

# chmod -R a+rwx /root/metastore_db
# chmod -R a+rwx /usr/lib/jvm/metastore_db
# chmod -R a+rwx /home/spark/metastore_db
```

步骤 04 设置 /usr/local/spark/conf 文件：

```
# cd /usr/local/spark/conf
# cp slaves.template slaves
# vi slaves
slave-1
```

```
slave-2
```

启动时只需在 192.168.244.151（master）上执行：

```
# su - spark （请使用 spark执行）
$ cd /usr/local/spark
$ sbin/start-all.sh
```

在 chrome 上执行 192.168.244.151:8080，结束 Standlonespark：

```
$ sbin/stop-all.sh
```

安装 Jupyter-Notebook

下载 Anaconda Python 2.7，地址为 https://www.continuum.io/downloads#_unix：

```
# cd /user/local
# wget http://repo.continuum.io/archive/Anaconda2-4.0.0-Linux-x86_64.sh
# bash Anaconda2-4.0.0-Linux-x86_64.sh
# vi /etc/profile
export PATH="/root/anaconda2/bin:$PATH"

# source /etc/profile
# jupyter notebook --generate-config
# cd
# cd .jupyter
# vi jupyter_notebook_config.py
```

❖ 设置 master

```
c.NotebookApp.ip = '192.168.244.151'
c.NotebookApp.open_browser = False
c.NotebookApp.port = 8888
```

❖ 设置 slave-1

```
c.NotebookApp.ip = '192.168.244.152'
c.NotebookApp.open_browser = False
c.NotebookApp.port = 8888
```

❖ 设置 slave-2

```
c.NotebookApp.ip = '192.168.244.153'
c.NotebookApp.open_browser = False
c.NotebookApp.port = 8888
```

❖ 安装相关程序包

```
# conda install matplotlib
# conda install pandas

# yum install -y python-pip
# pip install graphviz    （不能使用 conda）
```

```
# PYSPARK_DRIVER_PYTHON=ipython PYSPARK_DRIVER_PYTHON_OPTS="notebook --no- browser"
/usr/local/spark/bin/pyspark --master local[2]
```

❖ 启动浏览器

```
http://192.168.244.151:8888 (master)
http://192.168.244.152:8888 (slave-1)
http://192.168.244.153:8888 (slave-2)
```

❖ 离开 pyspark

```
$ ctrl +C
```

安装R于master、slave-1及slave-2

步骤01 安装 EPEL(Extra Packages for Enterprise Linux) Repository on CentOS 7.x。

```
# wget http://dl.fedoraproject.org/pub/epel/7/x86_64/e/epel-release-7-5.noarch.rpm

# rpm -ivh epel-release-7-5.noarch.rpm
Verify EPEL Installation
# yum repolist | grep "^epel\|repo id"

[root@localhost ~]# yum repolist | grep "^epel\|repo id"
repo id            repo name                                      status
epel/x86_64        Extra Packages for Enterprise Linux 7 - x86_64  6,158
[root@localhost ~]#
```

步骤02 安装 thrift-0.9.3。可到 http://thrift.apache.org/download 手动下载安装程序，再使用 winSCP 上传到 /usr/local。

❖ 安装步骤

（1）安装 boost-devel：

```
# yum install boost-devel
```

（2）安装 openssl-devel：

```
# yum install openssl-devel
# yum install automake libtool flex bison pkhonfig gcc-c++
```

（3）解压缩：

```
# tar zxvf thrift-0.9.3.tar.gz
```

（4）执行./configure（在 thrift-0.9.3 文件夹下），确认 Building C++ Library 为 yes：

```
# cd /usr/local/thrift-0.9.3
# ./configure
thrift 0.9.3
Building C++ Library ......... : yes
Building C (GLib) Library .... : no
```

```
Building Java Library ........ : no
Building C# Library ......... : no
Building Python Library ...... : yes
Building Ruby Library ........ : no
Building Haskell Library ..... : no
Building Perl Library ........ : no
Building PHP Library ......... : no
Building Erlang Library ...... : no
Building Go Library .......... : no
Building D Library ........... : no

C++ Library:
Build TZlibTransport ...... : yes
Build TNonblockingServer .. : no
Build TQTcpServer (Qt) .... : no

Python Library:
Using Python ............. : /usr/bin/python

If something is missing that you think should be present, please skim the output of configure
to find the missing component.  Details are present in config.log.
```

（5）修改/usr/local/thrift-0.9.3/lib/cpp/thrift.pc，将includedir=${prefix}/include/修改为includedir=${prefix}/include/thrift：

```
# vi /usr/local/thrift-0.9.3/lib/cpp/thrift.pc includedir=${prefix}/include/thrift
```

（6）到 thrift-0.9.3 目录中执行 make：

```
# cd /usr/local/thrift-0.9.3
# make
```

（7）make install，若无错误，即安装成功：

```
# make install
```

（8）设置 pkg_config_path：export PKG_CONFIG_PATH=/usr/local/lib/pkgconfig：

```
# export PKG_CONFIG_PATH=/usr/local/lib/pkgconfig
```

（9）安装完毕后执行 pkg-config --cflags thrift：

```
-I /usr/local/include/thrift
 pkg-config --cflags thrift
```

（10）复制 libthrift-0.9.3.so 文件到/usr/lib：

```
cp -Rp /usr/local/lib/libthrift-0.9.3.so /usr/lib
```

（11）载入 libthrift-0.9.3.so 文件：

```
ldconfig /usr/lib/libthrift-0.9.3.so
```

注意,若 thrift 因故未安装成功,最好 make clean 或重新解压缩后再执行 ./configure 步骤。

步骤 03 安装 R。

```
yum install R
```

步骤 04 执行 R 软件。

```
R
```

步骤 05 输入 q() 退出 R。

```
q()
```

步骤 06 安装 SparkR 2.1 程序包。

```
# R CMD INSTALL /usr/local/spark/R/lib/SparkR -l /usr/lib64/R/library/
```

步骤 07 安装 R-STUDIO。

下载 R-STUDIO Server 版:

```
# wget https://download2.rstudio.org/rstudio-server-rhel-1.0.136-x86_64.rpm

# yum install --nogpgcheck rstudio-server-rhel-1.0.136-x86_64.rpm

# service rstudio-server stop
# service rstudio-server start  (默认已启动)
```

步骤 08 使用 R-STUDIO。

在 Windows 下以 chrome 开启 http://192.168.244.151:8787

- http://192.168.244.152:8787
- http://192.168.244.153:8787

安装 Cassandra

Cassandra 为 NoSQL 的一种,用户可自行决定是否安装。

```
$ su -
# vi /etc/yum.repos.d/datastax.repo
[datastax-ddc]
name = DataStax Repo for Apache Cassandra
baseurl = http://rpm.datastax.com/datastax-ddc/3.2
enabled = 1
gpgcheck = 0


# yum install datastax-ddc


# vi /etc/cassandra/conf/cassandra.yaml
```

❖ 设置 master

```
cluster_name: 'CassandraCluster'
num_tokens: 256
seed_provider:
# Addresses of hosts that are deemed contact points.
# Cassandra nodes use this list of hosts to find each other and learn
# the topology of the ring.  You must change this if you are running
# multiple nodes!
- class_name: org.apache.cassandra.locator.SimpleSeedProvider
parameters:
# seeds is actually a comma-delimited list of addresses.
# Ex: "<ip1>,<ip2>,<ip3>"
- seeds: "192.168.244.151,192.168.244.152,192.168.244.153"

listen_address: 192.168.244.151
rpc_address: 192.168.244.151 endpoint_snitch: GossipingPropertyFileSnitch


# Whether to start the thrift rpc server.
start_rpc: true
```

❖ 设置 slave-1

```
cluster_name: 'CassandraCluster'
num_tokens: 256
seed_provider:
# Addresses of hosts that are deemed contact points.
# Cassandra nodes use this list of hosts to find each other and learn
# the topology of the ring.  You must change this if you are running
# multiple nodes!
- class_name: org.apache.cassandra.locator.SimpleSeedProvider
parameters:
# seeds is actually a comma-delimited list of addresses.
# Ex: "<ip1>,<ip2>,<ip3>"
- seeds: "192.168.244.151,192.168.244.152,192.168.244.153"

listen_address: 192.168.244.152
rpc_address: 192.168.244.152
endpoint_snitch: GossipingPropertyFileSnitch


# Whether to start the thrift rpc server.
start_rpc: true
```

❖ 设置 slave-2

```
cluster_name: 'CassandraCluster'
num_tokens: 256
seed_provider:
# Addresses of hosts that are deemed contact points.
# Cassandra nodes use this list of hosts to find each other and learn
# the topology of the ring.  You must change this if you are running
# multiple nodes!
```

```
    - class_name: org.apache.cassandra.locator.SimpleSeedProvider
      parameters:
      # seeds is actually a comma-delimited list of addresses.
      # Ex: "<ip1>,<ip2>,<ip3>"
        - seeds: "192.168.244.151,192.168.244.152,192.168.244.153"

    listen_address: 192.168.244.153
    rpc_address: 192.168.244.153
    endpoint_snitch: GossipingPropertyFileSnitch

    # Whether to start the thrift rpc server.
    start_rpc: true

    # vi /etc/cassandra/conf/cassandra-rackdc.properties
    dc=dc1
    rack=rack1
    # Add a suffix to a datacenter name. Used by the Ec2Snitch and Ec2MultiRegionSnitch
    # to append a string to the EC2 region name.
    #dc_suffix=
    # Uncomment the following line to make this snitch prefer the internal ip when possible,
as the Ec2MultiRegionSnitch does.
    prefer_local=true
    # vi /etc/profile

    export CQLSH_NO_BUNDLED=true
```

清除 Cassandra data（若有重新设定时）：

```
    # rm -rf /var/lib/cassandra/data/system/*
```

安装 Apache-Cassandra 与 R 的驱动程序：

```
    # cd /usr/local
    # wget http://www-eu.apache.org/dist/cassandra/3.7/apache-cassandra-3.7-bin.tar.gz
```

或

```
    # wget https://archive.apache.org/dist/cassandra/3.7/apache-cassandra-3.7-bin.tar.gz

    # tar zxvf apache-cassandra-3.7-bin.tar.gz

    # cd /usr/local/apache-cassandra-3.7/lib
    # wget https://storage.googleapis.com/google-code-archive-downloads/v2/apache-extras.org/cassandra-jdbc/cassandra-jdbc-1.2.5.jar

    # wget http://repo1.maven.org/maven2/org/apache/cassandra/cassandra-all/1.1.0/cassandra-all-1.1.0.jar

    # cd /usr/local
    # chown -R spark.spark apache-cassandra-3.7
```

安装 Apache-Cassandra 与 Spark 的驱动程序（可先到 https://spark-packages.org/package/datastax/spark-cassandra-connector 检查合适版本）。

```
# cd /home/spark

# wget http://dl.bintray.com/spark-packages/maven/datastax/spark-cassandra-connector/2.0.0-M2-s_2.11/spark-cassandra-connector-2.0.0-M2-s_2.11.jar
```

若无目录，则先创建 /home/spark/.m2/repository/com/datastax/spark/spark-cassandra-connector/2.0.0-M2-s_2.11/ 目录，再复制 spark-cassandra-connector-2.0.0-M2-s_2.11.jar。

```
# cp spark-cassandra-connector-2.0.0-M2-s_2.11.jar /home/spark/.m2/repository/com/datastax/spark/spark-cassandra-connector/2.0.0-M2-s_2.11/

# cp spark-cassandra-connector-2.0.0-M2-s_2.11.jar /usr/local/spark/spark-cassandra-connector-2.0.0-M2-s_2.11.jar
```

启动 Cassandra（master、slave-1 和 slave-2 都需启动）：

```
# su cassandra/usr/sbin/cassandra
```

检查 cassandra node status：

```
# nodetool status
```

测试 cqlsh 并退出：

```
master
# cqlsh 192.168.244.151 9042
exit

slave-1
# cqlsh 192.168.244.152 9042
exit

slave-2
# cqlsh 192.168.244.153 9042
exit
```

❖ 需要 Debug CQL 时

cqlsh 版本不对，检查路径：

```
# which cqlsh
/root/anaconda2/bin/cqlsh

# ls -l /usr/bin/cqlsh
-rwxr-xr-x. 1 root root 834 Jan 20 08:30 /usr/bin/cqlsh
```

以 /usr/bin/cqlsh 取代 /root/anaconda2/bin/cqlsh：

```
# mv /root/anaconda2/bin/cqlsh /root/anaconda2/bin/cqlsh.o
# ln -s /usr/bin/cqlsh /root/anaconda2/bin/cqlsh
```

```
No module named cqlshlib
# find / -name cqlshlib
/usr/lib/python2.7/site-packages/cqlshlib

# cp /usr/lib/python2.7/site-packages/cqlshlib /usr/bin/

# cqlsh 192.168.244.151 9042
exit
```

❖ 安装 Python-Cassandra Driver

```
# pip install cassandra-driver
```

使用 cassandra CQL 指令进行数据的导出/导入。

❖ 在 master 上执行

```
# cqlsh 192.168.244.151 9042
```

（1）列出有哪些 keyspace：

```
DESCRIBE keyspaces;    // Output the names of all keyspaces.
```

（2）使用 keyspace：

```
CREATE KEYSPACE mykeyspace
WITH REPLICATION = { 'class' : 'SimpleStrategy', 'replication_factor' : 1 };

USE mykeyspace;
```

（3）创建 TABLE：

```
CREATE TABLE  Student (
NO    VARCHAR PRIMARY KEY,
NAME TEXT,
CLASS   VARCHAR
);
```

（4）输入数据：

```
INSERT INTO Student (no,name,CLASS) VALUES('B0001006', 'Frodo', 'IM1A');
```

（5）批次输入指令，类似 SQL STOREPROCEDURE：

```
BEGIN BATCH
INSERT INTO Student (no,name,class) VALUES('B0001007', 'Brodo', 'IM2A');
INSERT INTO Student (no,name,class) VALUES('B0001008', 'Krodo', 'IM3A');
INSERT INTO Student (no,name,class) VALUES('C0001018', '林大力 ', 'IM3A') ;
-->不能输入中文  APPLY BATCH;
```

（6）导出数据到 CSV 文件：

```
COPY Student (no,class,name)   TO 'temp_Student.csv';
```

（7）清空数据：

```
TRUNCATE Student;
```

（8）导入数据：

```
COPY Student (no,name,class) FROM 'temp_Student.csv';
```

（9）测试 Python-Cassandra(jupyter)：

```
from cassandra.cluster import Cluster
cluster = cluster(['192.168.244.151'])
session = cluster.connect()
session.set_keyspace('mykeyspace')
rows = session.execute('SELECT * FROM Student;')
for user_row in rows:
print user_row.no, user_row.name;
```

❖ 测试 Cluster 是否正常

在 slave-1 和 salve-2 上执行：

```
USE mykeyspace; Select * from Student;
```

启动方式

启动 master、slave-1 及 slave-2 的 Cassandra。

```
# su cassandra   /usr/sbin/cassandra
# nodetool enablethrift
```

启动 Standalone Spark，只需在 192.168.244.151 (master) 上执行。

```
# su - spark (请使用 spark 执行 )
$ cd /usr/local/spark
$ sbin/start-all.sh

$ sbin/stop-all.sh (结束时 )
```

Web 界面：

```
http://192.168.244.151:8888 (PYSPARK)
http://192.168.244.151:4040 (Spark)
http://192.168.244.151:8080 (Standalone Spark)
http://192.168.244.151:8787 (R)
```

参考文献

[1] Fayyad, Usama, Gregory Piatetsky-Shapiro, and Padhraic Smyth. "From data mining to knowledge discovery in databases." AI magazine 17.3 (1996): 37.

[2] Berry, Michael J., and Gordon Linoff. Data mining techniques: for marketing, sales, and customer support. John Wiley & Sons, Inc., 1997.

[3] Breiman, Leo, et al. Classification and regression trees. CRC press, 1984.

[4] Breiman, Leo. "Random forests." Machine learning 45.1 (2001): 5-32.

[5] Quinlan, J. Ross. C4. 5: programs for machine learning. Elsevier, 2014.

[6] Quinlan, J. Ross. "Induction of decision trees." Machine learning 1.1 (1986): 81-106.

[7] Vapnik, Vladimir Naumovich, and Vlamimir Vapnik. Statistical learning theory. Vol. 1. New York: Wiley, 1998.

[8] Cortes, Corinna, and Vladimir Vapnik. "Support-vector networks." Machine learning 20.3 (1995): 273-297.

[9] Wang, Sun-Chong. "Artificial neural network." Interdisciplinary Computing in Java Programming. Springer US, 2003. 81-100.

[10] Hagan, Martin T., Howard B. Demuth, and Mark H. Beale. Neural network design. Boston: Pws Pub, 1996.

[11] Nilsson, Nils J. "Learning machines." (1965).

[12] MacQueen, James. "Some methods for classification and analysis of multivariate observations." Proceedings of the fifth Berkeley symposium on mathematical statistics and probability. Vol. 1. No. 14.1967.

[13] Bezdek, James Christian. "Fuzzy mathematics in pattern classification." (1973).

[14] Holland, John L. Making vocational choices: A theory of careers. Prentice Hall, 1973.

[15] Holland, John H. Adaptation in natural and artificial systems: an introductory analysis with applications to biology, control, and artificial intelligence. U Michigan Press, 1975.

[16] Karaboga, Dervis. An idea based on honey bee swarm for numerical optimization. Vol. 200. Technical report-tr06, Erciyes university, engineering faculty, computer engineering department, 2005.

[17] Karaboga, Dervis, and Bahriye Basturk. "A powerful and efficient algorithm for numerical function optimization: artificial bee colony (ABC) algorithm." Journal of global optimization 39.3 (2007): 459- 471.

[18] Agrawal, Rakesh, and Ramakrishnan Srikant. "Fast algorithms for mining association rules."

Proc. 20th int. conf. very large data bases, VLDB. Vol. 1215. 1994.

[19] Agrawal, Rakesh, Tomasz Imieliński, and Arun Swami. "Mining association rules between sets of items in large databases." ACM SIGMOD Record. Vol. 22. No. 2. ACM, 1993.

[20] Zaki, Mohammed Javeed. "Scalable algorithms for association mining."Knowledge and Data Engineering, IEEE Transactions on 12.3 (2000): 372-390.

[21] Zaki, Mohammed Javeed, et al. "New Algorithms for Fast Discovery of Association Rules." KDD. Vol. 97. 1997.

[22] Luis Torgo, Data Mining with R: Learning with Case Studies, Chapman & Hall/CRC Press, 2011.

[23] Yangchang Zhao, R and data mining: examples and case studies, Academic Press, 2013.

[24] Graham Williams, Data mining with Rattle and R: the art of excavating data for knowledge discovery, Springer, 2011.

[25] Zhengxin Chen, Data mining and uncertain reasoning: an integrated approach, Wiley, 2001.

[26] Yanchang Zhao and Yonghua Cen, Data mining applications with R, Elsevier, 2014.

[27] Zhi-Hua Zhou, Ensemble methods: foundations and algorithms, Taylor & Francis, 2014.

[28] Vignesh Prajapati, Big data analytics with R and Hadoop: set up an integrated infrastructure of R and Hadoop to turn your data analytics into Big Data analytics, Packt Publishing, 2013.

[29] Vijay Srinivas Agneeswaran, Big data analytics beyond Hadoop: real-time applications with storm, spark, and more Hadoop alternatives, Pearson Education, 2014.

[30] Ethem Alpaydin, Introduction to machine learning, MIT Press, 2004.

[31] John J. Grefenstette, Genetic algorithms for machine learning, Kluwer Academic Publishers, 1994.

[32] Zne-Jung Lee, Chou-Yuan Lee, Wei-Feng Ling, and Yuan-Chih Lee, Apply Data Mining to Analyze the Rainfall of Landslide, IMETI 2016.

[33] http://spark.apache.org/.